MW01050012
AGNÈS KEFELI
From
LUCY
to COLUMBUS
REVISED EDITION

Kendall Hunt
publishing company

Kendall Hunt
publishing company
www.kendallhunt.com
Send all inquiries to:
4050 Westmark Drive
Dubuque, IA 52004-1840

ISBN 978-1-5249-4827-6

Published in the United States of America

Contents

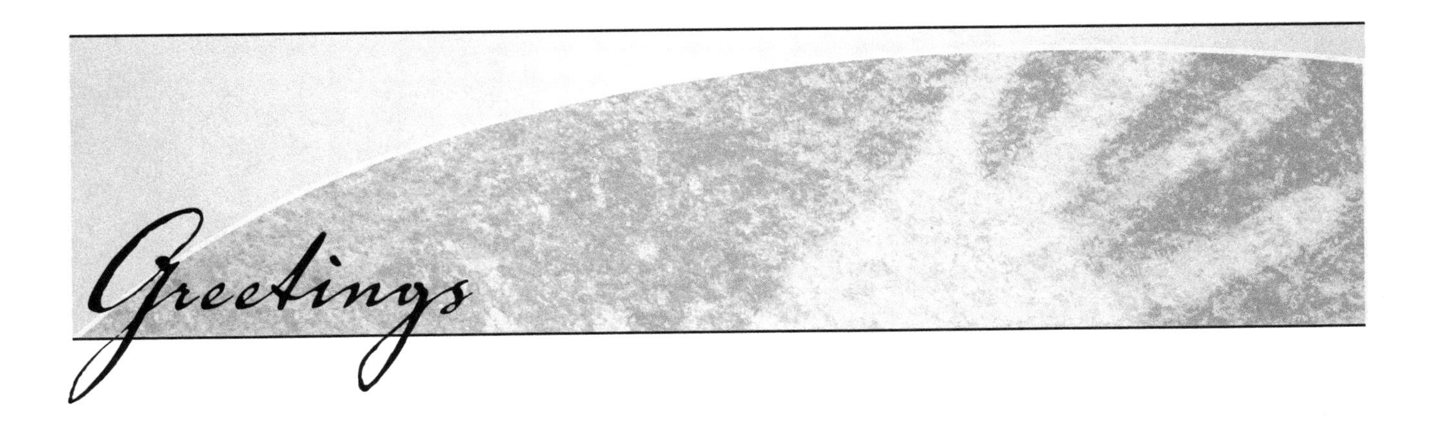

Dear Students:

Welcome to Global History!

This course will introduce you to some of the most important ideas and philosophies that have shaped our understanding of what it means to be human. We will travel through time from the prehistoric era to the great maritime revolution that united the Eastern and Western hemispheres for the first time in the fifteenth and sixteenth centuries. You will discover stunning works of art, such as the cave paintings in Spain, Australia, and Southern France; the Egyptian Book of the Dead; the great Gothic cathedrals of France; the paintings of the Italian and Northern Renaissance; the astonishing mud mosques of Black Africa; and the miniatures of Ottoman Turkey. You will live the lives of nomadic herders in the Eurasian steppes, travel by sea, and discover new worlds shaped by different ethical and legal codes.

At the end of the semester, you will understand the role that ecology, geography, and biology played, and still play, in human history and appreciate the importance of religion and philosophy in individual and collective life. Most important, you will understand the historical contexts behind many of today's current events and issues (e.g., the impact of religion and migration, the origins of our democratic values, and the history of various technologies that we take for granted today).

This workbook is divided into eighteen chapters. There you will find PowerPoint outlines, maps, crosswords, matching games, in-class exercises, questions on documentary films, and newspaper articles. Please bring your workbook every time you come to class.

Always check Blackboard for your weekly assignments and complementary materials (weekly messages, lectures in full sentences, and study guides for your exams).

Enjoy this trip around the world!

Dr. Agnès Kefeli
Principal Lecturer
Islamic Studies Certificate Coordinator
School of Historical, Philosophical, and Religious Studies
Arizona State University

BEFORE COMMON ERA
BCE

COMMON ERA
CE

6 5 4 3 2 1 . 1 2 3 4 5 6

Birth of Jesus of Nazareth

© dean bertoncelj/Shutterstock.com

Gorilla Using a Branch as a Tool

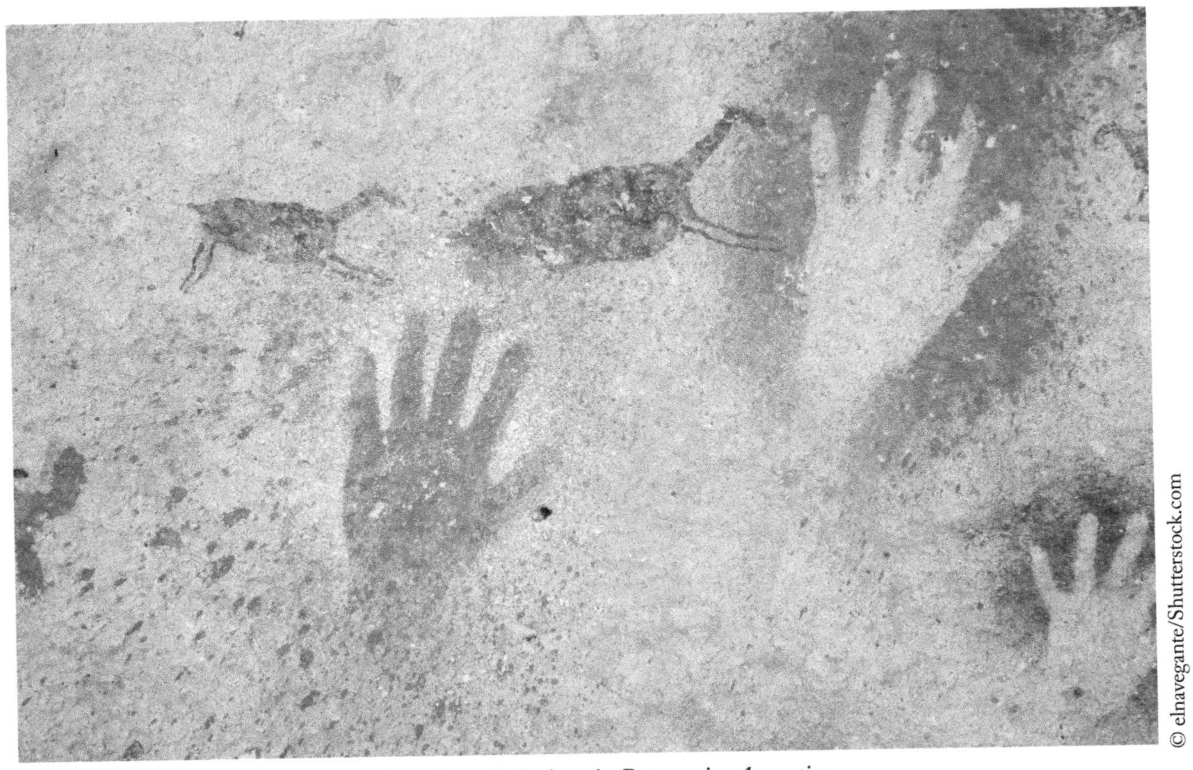

© elnavegante/Shutterstock.com

Ancient Cave Paintings in Patagonia, Argentina

FROM HUMAN ORIGINS TO AGRICULTURAL TRANSFORMATION (FROM FORAGING ECONOMY TO AGRICULTURE)

BEFORE 20th C., HOW DID WE UNDERSTAND HUMAN ORIGINS?

- Before Charles Darwin (19th c.)
  - In Europe, people turned to the Bible: God created each species individually
  - Each species had fixed characteristics
  - Human species and animals are separate
- After Charles Darwin
  - Nature is not fixed (volcanic action, erosion, earthquakes)
  - All species undergo gradual but constant change

ARE WE THAT DIFFERENT FROM ANIMALS?

- Not much, only 1.6 percent difference between human and chimpanzee DNA
- People used to say that animals have no culture
  - Culture = behaviors transmitted by learning, not just sprung from instinct
- Wrong, animals have culture (Jane Goodall)
  - Monkeys can make tools and teach their young to use them
  - Animal societies can be divided by clans/gangs. Clans fight for power and control
  - Animal societies are not rigid. They adapt to new situations
  - Animals have emotions

NEWS FLASH

- We are not the first who made tools
- We are not the first who taught our young
- We are not the first who made weapons
- We are not the first who stacked the bones of our dead
- Are we chimps then? How did we become human? What does it mean to be human?

ARE WE CHIMPS THEN?

- We descend from Hominids
- Hominids = early humanlike primates, genetically splitting from the chimpanzees 5 million years ago in Africa

LUCY = THE MOST FAMOUS HOMINID (AUSTRALOPITHECUS)

4 million to 1 million years ago in Eastern and Southern Africa

- Short stature and small brain size
- Walked first upright on trees, then on ground
- Omnivore
- Used sharp stone tools to remove meat from animals killed by beasts of prey

HOMO ERECTUS = 1 MILLION YEARS AGO

- Was fully stabilized on feet (no more climbing)
- Walked longer distances through a wide variety of habitats
- Had a bigger brain
- Fashioned more sophisticated tools (hand axes)
- Learned to start and attend fire
- Formed camp groups at cave entrances
- Hunted animals
- Migrated to North Africa + Europe + Asia (all temperate zones of eastern atmosphere)

WHY DID HOMINIDS SPLIT FROM APES?

- Ecology + adaptation to new environment (climate changes)
- Bipedalism (upright walking) [apes continued knuckle-walking]
- Some hominids lived on borderland of diverse environments (tropical rainforest, savannas, and grassland)
- In the less dense savanna tree stands, hominids started branch-walking on straight legs to reach farther

OUR COMMON ANCESTOR: HOMO SAPIENS = "CONSCIOUSLY THINKING MAN"

- Appeared in East Africa
- Large brain well developed in the frontal region where conscious and reflective thought takes place
- Homo sapiens developed a throat with vocal cords, which enabled them to enunciate hundreds of distinct sounds
- Migrated to the temperate lands of Africa, Europe, and Asia; there they encountered Homo erectus, but soon migrated to colder regions

HOMO SAPIENS = FORAGERS

- Nomadic life: their movement coincided with animal seasonal migrations and life cycles of plants
- Survived Ice Age (150,000 to 20,000 years ago) and exploited it for migration from Siberia to the Americas
    - Ice Age = The earth tilts on its axis every 100,000 years and Northern Hemisphere moves away from sun
- Fashioned warm clothes from animal skins (sewing needles from animal bones)
- Hunters (first spears, knives, bows, arrows, hooks for catching fish)
    - Mammoths, woolly rhinoceros, giant kangaroos became extinct
- Sheltered in caves AND huts made from branches, bones, and animal skins
- Domesticated dogs + built fires at will + stored food + invented canoe

ANOTHER GROUP IN EUROPE + MIDDLE EAST: NEANDERTHAL (EARLIEST BURIAL PLACES)

- When Homo sapiens reached Europe another group was already in place, the Neanderthal
- Neanderthal = successor to Homo erectus? (unclear)
- Evidence of interbreeding and trading between Homo sapiens and Neanderthals
- Neanderthals reflected on life and death (earliest burial places)

- Homo sapiens were superior to Neanderthals because of their superior linguistic abilities (different throat structures)
- Neanderthals did not survive Ice Age

WHEN DID HOMO SAPIENS AND NEANDERTHALS LEAVE THE WORLD OF ANIMALS?

- When they started thinking in abstract/symbolic terms (and not solely in practical terms)
- Earliest evidence of abstract thinking came from
  - Neanderthal burial places: near Baghdad (Iraq), the dead were laid on a bed of wildflowers
  - Homo sapien jewelry found in burial places
- Jewelry not practical tools but symbols of feelings

AFTER ICE AGE: BEGINNING OF AGRICULTURE

- Neolithic era = new stone age [roughly 12,000 to 6,000 years ago]
- Polished stone tools
- People moved away from foraging to cultivation
- Agriculture was impossible in cold places
- More stable climactic conditions + rainfalls
- No need to migrate as much

AGRICULTURAL REVOLUTION

- Agricultural revolution = transition from collecting food in the wild to domesticating and producing particular plants and animals for human consumption
- Why? Change of climate (big thaw)
- Who prepared for it? Gender and agriculture
  - Women began nurturing plants instead of simply collecting food in the wild
  - Men began capturing animals and domesticating them (by giving them food and supervising their breeding)
- Agriculture emerged in several different places independently
- Migrants and traders brought new seeds and techniques with them

MAIN CONSEQUENCES: BEGINNING OF URBANIZATION + PATRIARCHY

- More food supplies meant people were better fed and could trade surpluses
- Population explosion (need of labor for fields)
- Adoption of new forms of social organizations
  - First people settled near their fields in permanent villages
  - Villages became towns
  - Need to build walls to defend against predators + need for political centralization to organize labor force (irrigation, canals, etc.)
- Labor specialization: pottery (food containers), metallurgy (copper for jewels, knives, axes, hoes, and weapons), and textile from strains of plants and animals
- Social distinctions, inequality, and patriarchy
  - In foraging communities, people could not accumulate wealth
  - It became important to keep the land in same families' hands to consolidate wealth (need to control paternity for inheritance, hence women's movements)

What scientific disciplines do the specialists in the videos you watched use to extract information about the prehistoric past?

1.

2.

3.

4.

PREHISTORIC ART

What did you learn today?

1.

2.

3.

4.

ÇATAL HÖYÜK

1. TIME

2. PLACE

3. SIGNIFICANCE

PREHISTORY

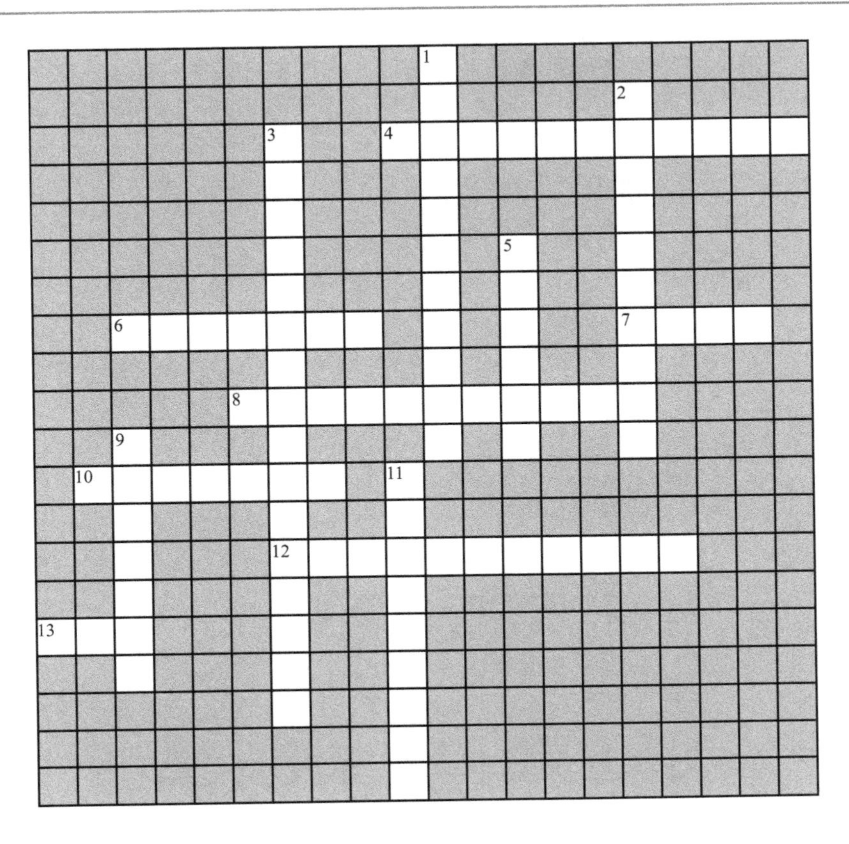

ACROSS

4. Old Stone Age (earliest and longest stage of cultural development, roughly 2 million BCE–10,000 BCE)
6. Last name of a British ethologist who argues that animals have culture (21th c.)
7. The most famous 3.2-million-year-old Australopithecus found in Ethiopia by ASU professor, Donald Johanson in 1974
8. "Wise human" in Latin or ancestor of all modern people, superior to other hominids because of his/her superior linguistic abilities
10. Humans and two-legged pre-human predecessors who genetically split from the chimpanzees
12. First hominid that left Africa and adapted to many different climactic zones in Europe and Asia
13. The ____ Age refers to a stretch of time when huge glaciers spread across most of the globe

DOWN

1. Group of large-brained hominids discovered in Germany and known for laying their dead on beds of wildflowers
2. Upright walking (first human characteristic)
3. Pre-human and omnivore species that existed before Homo Erectus and could walk upright on two legs
5. Last name of a famous nineteenth-century British naturalist and geologist who hypothesized the existence of a common ancestor for great apes and humans
9. Subsists by gathering wild plants and hunting animals
11. New Stone Age which marks the beginning of agriculture and herding (roughly 10,000 to 4,000 BCE)

Article One

Who Apes Whom?

BY FRANS DE WAAL

ATLANTA--When I learned last week about the discovery of an early human relative deep in a cave in South Africa, I had many questions. Obviously, they had dug up a fellow primate, but of what kind?

The fabulous find, named Homo naledi, has rightly been celebrated for both the number of fossils and their completeness. It has australopithecine-like hips and an ape-size brain, yet its feet and teeth are typical of the genus Homo.

The mixed features of these prehistoric remains upset the received human origin story, according to which bipedalism ushered in technology, dietary change and high intelligence. Part of the new species' physique lags behind this scenario, while another part is ahead. It is aptly called a mosaic species.

We like the new better than the old, though, and treat every fossil as if it must fit somewhere on a timeline leading to the crown of creation. Chris Stringer, a prominent British paleoanthropologist who was not involved in the study, told BBC News: "What we are seeing is more and more species of creatures that suggests that nature was experimenting with how to evolve humans, thus giving rise to several different types of humanlike creatures originating in parallel in different parts of Africa."

This represents a shockingly teleological view, as if natural selection is seeking certain outcomes, which it is not. It doesn't do so any more than a river seeks to reach the ocean. News reports spoke of a "new ancestor," even a "new human species," assuming a ladder heading our way, whereas what we are actually facing when we investigate our ancestry is a tangle of branches. There is no good reason to put Homo naledi on the branch that produced us. Nor does this make the discovery any less interesting.

Every species in our lineage tells us something about ourselves, because the hominoids (humans, apes and everything in between) are genetically extremely tight. We have had far less time to diverge than the members of many other animal families, like the equids (horses, zebras, donkeys) or canids (wolves, dogs, jackals). If it hadn't been for the human ego, taxonomists would long ago have squeezed all hominoids into a single genus.

The standard story is that our ancestors first left the apes behind to become australopithecines, which grew more sophisticated and brainier to become us. But what if these stages were genetically mixed up? Some scientists have claimed early hybridization between human and ape DNA. Did our ancestors, after having split off, keep returning to the apes in the same way that today's grizzlies and polar bears still interbreed occasionally?

Instead of looking forward to a glorious future, our lineage may have remained addicted to the hairy embrace of its progenitors. Other scientists, however, keep sex out of it and speak of incomplete lineage separation. Either way, our heritages are closely intertwined.

The problem is that we keep assuming that there is a point at which we became human. This is about as unlikely as there being a precise wavelength at which the color spectrum turns from orange into red. The typical proposition of how this happened is that of a mental breakthrough — a miraculous spark — that made us radically different. But if we have learned anything from more than 50 years of research on chimpanzees and other intelligent animals, it is that the wall between human and animal cognition is like a Swiss cheese.

Apart from our language capacity, no uniqueness claim has survived unmodified for more than a decade since it was made. You name it — tool use, tool making, culture, food sharing, theory of mind, planning, empathy, inferential reasoning — it has all been observed in wild primates or, better yet, many of these capacities have been demonstrated in carefully controlled experiments.

We know, for example, that apes plan ahead. They carry tools over long distances to places where they use them, sometimes up to five different sticks and twigs to raid a bee nest or probe for underground ants. In the lab, they fabricate tools in anticipation of future use. Animals think without words, as do we most of the time.

Undeterred by Homo naledi's relatively small brain, however, the research team sought to stress its humanity by pointing at the bodies in the cave. But if taking this tack implies that only humans mourn their dead, the distinction with apes is being drawn far too sharply.

Apes appear to be deeply affected by the loss of others to the point of going totally silent, seeking comfort from bystanders and going into a funk during which they don't eat for days. They may not inter their dead, but they do seem to understand death's irreversibility. After having stared for a long time at a lifeless companion — sometimes grooming or trying to revive him or her — apes move on.
Since they never stay in one place for long, they have no reason to cover or bury a corpse. Were they to live in a cave or settlement, however, they might notice that carrion attracts scavengers, some of which are formidable predators, like hyenas. It would absolutely not exceed the ape's mental capacity to solve this problem by either covering odorous corpses or moving them out of the way.

The suggestion by some scholars that this requires belief in an afterlife is pure speculation. We simply don't know if Homo naledi buried corpses with care and concern or unceremoniously dumped them into a faraway cave to get rid of them.

It is an odd coincidence that "naledi" is an anagram of "denial." We are trying way too hard to deny that we are modified apes. The discovery of these fossils is a major paleontological breakthrough. Why not seize this moment to overcome our anthropocentrism and recognize the fuzziness of the distinctions within our extended family? We are one rich collection of mosaics, not only genetically and anatomically, but also mentally.

How does article 1 change your view of human origins?

Chapter 2

Early Agricultural Centers and Urbanization

© Ialeks/Shutterstock.com

Noah's Ark

Part 1: Mesopotamia And Palestine

Mesopotamia on the Globe

© AridOcean/Shutterstock.com

Tigris and Euphrates in Iraq Today

© pavalena/Shutterstock.com

© Michael Rosskothen/Shutterstock.com

A Ziggurat

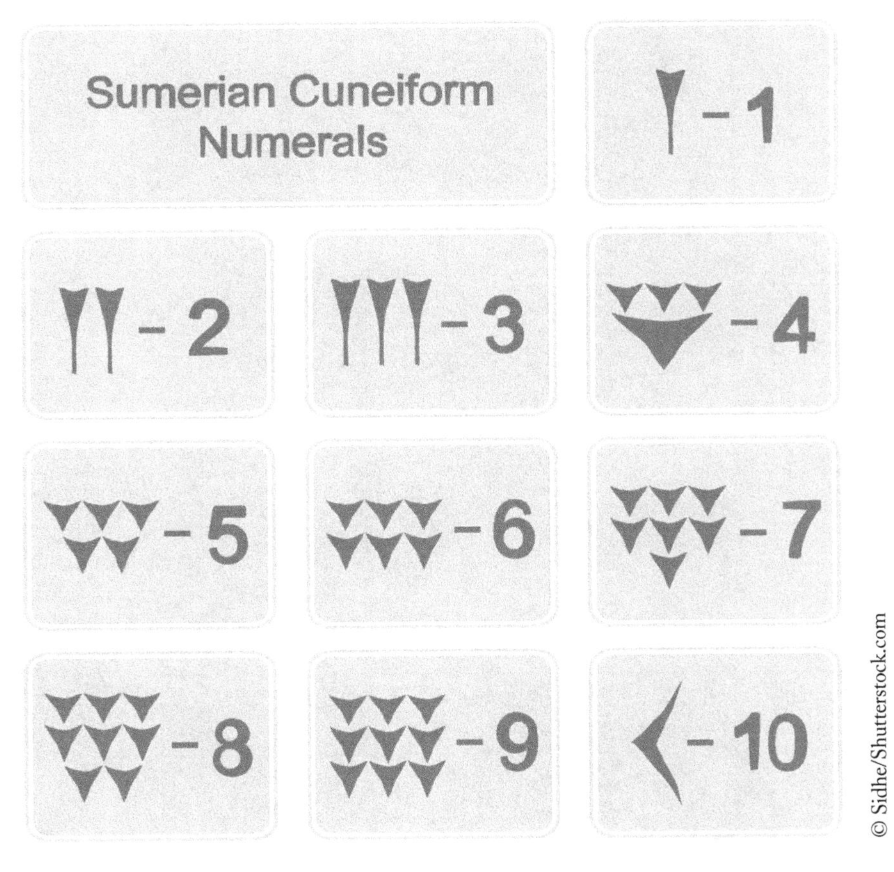

© Sidhe/Shutterstock.com

Cuneiform Numerals

GILGAMESH

© Fedor Selivanov/Shutterstock.com

Gilgamesh, Fifth King of Sumerian City Uruk

Significance: What can historians learn from the myths of Mesopotamia?

MESOPOTAMIA AND PALESTINE

VERY FIRST AGRARIAN CENTERS

- Were located in Middle East, India, and China
- Roughly 4,000/3,000 years ago
- Developed along rivers
- Were not isolated from each other
- How? Nomadic people (herders) from Europe and Central Asia connected them through migrations, trade, and conquests

MESOPOTAMIA

- Very first agrarian center
- Means the land between two rivers: Tigris and Euphrates
- Region known for its floods
- Only irrigation allowed farming + irrigation produced twice as much wheat or barley as a rain-fed field
- Produced the world's first cities/first city-states
  - City-states included surroundings (countryside)

GILGAMESH

- Fifth King of Sumerian city Uruk, located on Euphrates River
- Later remembered as a god
- Gilgamesh's stories recorded on clay tablets (oldest epic poem in history)
- Gilgamesh represents the city
- He ordered building of defensive walls and temples in Uruk

URUK, SUMERIAN CITY

- Big city-state (up to 80,000 inhabitants) in southern part of Mesopotamia
- First plow was found
- Invented potter's wheel (pots helped store food)

GILGAMESH EPIC

- Is a myth
- Myths = symbolic foundational stories about the origins and destiny of human beings and their world
- Myths express worldview/religious beliefs

MYTHS ANSWER IMPORTANT QUESTIONS

- How did the universe come into existence?
- What should be our relationship with the world of nature?
- Why do human beings exist?
- Why is there suffering in the world?
- How should we deal with it?
- What happens when we die?
- What is important for the people who produced these myths?

DO MYTHS HAVE A HISTORICAL VALUE?

YES, indeed

- Gilgamesh myth tells us that
  - Human beings struggled with nature (initially floods made farming difficult)—people saw themselves at the mercy of the gods of nature
  - There was interaction between Sumeria and ancient Hebrews (migrant Semitic people) [see Noah's flood]
- For very long, biblical stories were considered unique

HOW DID MESOPOTAMIAN CITIES DIFFER FROM NEOLITHIC SETTLEMENTS?

- Places of more than 5,000 people (Uruk)
- Centers of political and military authority + economic trade
- Jurisdiction extended into surrounding regions
- Massive buildings
- Landowning priests maintained organized religions
- Scribes developed traditions of writing and formal education

NEW INVENTIONS

- Plow
- Potter's wheel (easier to make jars, pots, bowls)
- Cart wheel
- Bronze (alloy of copper and arsenic or tin) metallurgy [bronze knives and bronze-tipped plows]
- Shipbuilding (allowed trade with India)

WRITING + CALENDAR

- Cuneiform writing on clay tablets with stylus
- Cuneiform comes from Latin = wedge shaped
- Symbols represented sounds, syllables, ideas, and objects
- Pictograph: a word-picture giving a stylized, standardized representation of an object
- Calendar (divided year into 12 months and hours of the day into 60 minutes, each composed of 60 seconds)

ZIGGURATS

- Stepped pyramids that housed temples and altars to principal local deity (built at the city center)
- Did they influence Babel story in Bible?
- In Uruk, one temple was devoted to Goddess Inanna (later known as Ishtar), goddess of love, fertility, and war
- Temples functioned as banks, helped those in need, supplied grain in case of famine, and provided ransoms

FIRST EMPIRE (SARGON OF AKKAD): NEW FORM OF ORGANIZATION

- After Sumerians, Akkadians (Semitic people) took over
- Sargon of Akkad (24th c. BCE) [Creator of first empire]

- Legend says: Sargon was abandoned by mother who put him in basket and sent him adrift on river (= biblical story of Moses)
- Sargon took over city-states; his empire embraced all of Mesopotamia
- Sargon's daughter, high priestess of goddess Inanna(= Ishtar) + first known female writer

BABYLONIAN HAMMURABI (18th c.)

- Relied on centralized bureaucratic rule and regular taxation through deputies stationed in the territories he controlled
- Sargon used to go from one city to another and have his army fed by the population, which was not popular
- Hammurabi is known for compiling the most extensive Mesopotamian law code

THE BABYLONIAN HAMMURABI'S LAW CODE (BEFORE MOSES)

- Proclaimed that gods had chosen Hammurabi to rule and uphold justice
- Prescribed death for murder, theft, fraud, false accusations, sheltering runaway slaves, adultery for women
- Civil laws regulated prices, wages, commercial dealings, marital relationships, conditions of slavery
- Preceded the Hebrew Mosaic tablets

GENERAL PRINCIPLES

- Protection of the weak (orphans, widows)
- Slavery for debts cannot last more than three years
- A child given away as apprentice for debt repayment cannot be mistreated
- No incest
- *Lex Talionis* (= law of retaliation): an eye for an eye
- However, *Lex Talionis* did not apply to nobles who could get away with a fine. No corporal punishment.
- Men = heads of household but women were protected from unreasonable treatment by husband

MEN IN HAMMURABI'S CODE

- Men = heads of household
- Could sell wife and children into slavery to satisfy debts
- Adulterous wives were drowned but men could engage in consensual sexual relationships with other women without penalty
- However, men's power was also limited
  - Could not disown own children (unless children committed something really wrong twice)
  - Could not repudiate wife if sterile without some form of compensation

WOMEN IN HAMMURABI'S CODE

- If a man rapes a virgin wife or child-wife, he should be killed
- A woman could go back to her father's house with her dowry if her husband was not congenial to her
- Women had property rights (they could buy, sell, inherit, and pass to descendants)

FAMILY LAW IN ISRAEL AND BABYLONIA

- *MOSAIC LAW IN DEUTERONOMY*
  - Only man can initiate divorce (Deuteronomy 24:1-4)
  - Children who oppose parents are put to death (Deuteronomy 21:18-21)
- *HAMMURABI'S CODE*
  - A woman can initiate divorce if her husband neglects her (and if she is known for being a perfect housewife) + she can take dowry back and go back to father
  - Children could be disowned but not put to death

ASSYRIANS: FIRST PROFESSIONAL ARMY

- Built first professional army divided into standardized units + adorned palaces with battle scenes
- Officers appointed on basis of merit
- Supplemented infantry with cavalry (horse-drawn chariots)
- Used recently invented iron weapons
- Deported the ten northern tribes of Israel
- Maintained Hammurabi's law code + large libraries and preserved famous Epic of Gilgamesh

NEBUCHADNEZZAR AND THE NEW BABYLONIAN EMPIRE (6TH C. BCE)

- Known for destroying Solomon's temple in Jerusalem and deporting Jews to Babylonia
- Main consequences of Babylonian exile
  - Diaspora
  - Synagogues
  - Babylon was famous for luxurious hanging gardens

MESOPOTAMIAN INFLUENCE ON HEBREWS/JEWS/ISRAELITES

- In Genesis (first book of the Torah in Hebrew Bible)
  - Abraham came from Sumerian city of Ur and then migrated to northern Mesopotamia
  - Noah's flood = variation of many Mesopotamian flood stories
  - Tower of Babel = a *ziggurat*? (symbol of false Mesopotamian gods)
- In Exodus, story of the infant Moses placed by his mother in a basket = Akkaddian Sargon's story
- In development of biblical law (notion of *Lex Talionis*)
- In biblical recognition of Mesopotamian gods (even though descendants of Abraham pledged allegiance and loyalty to their god, Yahweh)
- In creation of a Mesopotamian-style monarchy (Saul, David, and Solomon) [11-10th c. BCE]
- There is concrete evidence that Jerusalem traded with Mesopotamia

BIG DIFFERENCES

- *MESOPOTAMIAN WORLDVIEW*
  - Polytheistic (among gods, Marduk, patron deity of city of Babylon)
  - Gods portrayed in human form
  - Gods are vengeful and unfair (reflect harshness of ecology)
- *ANCIENT JUDAISM*
  - Moved away from polytheism to henotheism, then monotheism
  - Yahweh was not visible

- Yahweh is personal, omnipotent, omniscient, immortal, transcendent, and just
- Yahweh is not vengeful but forgiving even when his people occasionally turn away from him

ANCIENT JUDAISM: THREE COVENANTS (PACTS)

- Noah
- Abraham
  - Promised to worship God alone
  - God promised Abraham
    - Land of Canaan
    - Descendants
- Mosaic covenant extended Abrahamic covenant
  - God gave a systematic law at Sinai (Mosaic tablets)

FROM HENOTHEISM TO MONOTHEISM

- Yahweh was Abraham's tribal god or protector of his clan (henotheism)
- After Moses, biblical text says there is only one god (but nothing is said yet that all other gods are false)
- Prophets (Isaiah) declared that all other gods are false
  - Prophets = God's messengers
  - Prophets = interpreters of major traumatic historical events (Assyrians' deportation of ten northern tribes and later Babylonians' deportation of Jews to Babylonia)

DEFINITIONS

- Polytheism = belief in many gods
- Henotheism = temporary elevation of one of many gods to the highest rank that can be accorded, verbally or ritualistically
- Monotheism = belief in one god only (Prophets: God of all nations)

BRIEF CHRONOLOGY OF JEWISH HISTORY

- Moses
- Davidic monarchy (Saul, David, Solomon)
- Temple completed under Solomon
- After Solomon's death (division of kingdom) [Judea south + northern kingdom]
- Assyrians
- Babylonians (destruction of kingdom + exile)
- Return (Cyrus)

WHAT DID THE SUMERIANS INVENT?

1.

2.

3.

4.

5.

PLACE THESE DYNASTIES IN CHRONOLOGICAL ORDER

1. Akkadians •	• a. 1
2. Babylonians •	• b. 2
3. Neo-Babylonians •	• c. 3
4. Sumerians •	• d. 4
5. Assyrians •	• e. 5

Please format your answers as (1, a) below

WHO DID WHAT?

1. Gilgamesh •	• a. Destroyed first temple in Jerusalem
2. Sargon of Akkad •	• b. Created most extensive law code in Mesopotamia
3. Hammurabi •	• c. Created first empire in Mesopotamia
4. Nebuchadnezzar •	• d. Built defensive walls and temples in Uruk (Sumeria)
5. Moses •	• e. Deported the ten northern tribes
6. Assyrians •	• f. Received the Ten Commandments on Mount Sinai

Please format your answers as (1, a) below

MESOPOTAMIA

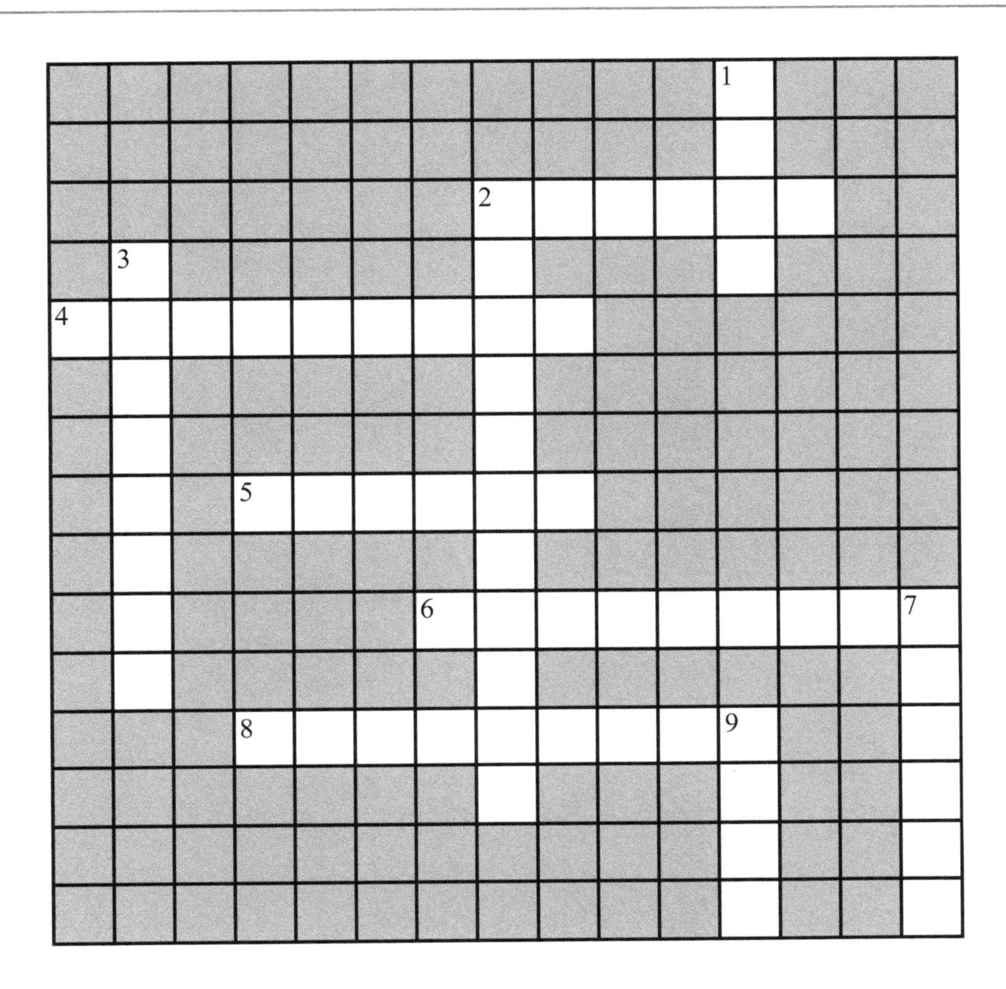

ACROSS

2. Patron-god of the city of Babylon and one of the most important gods of Mesopotamia. He is often represented as (or with) a snake-dragon
4. Fifth Sumerian king who built defensive walls and temples in Uruk
5. Created first empire in Mesopotamia
6. Famous Babylonian king who compiled a law code inscribed on a black stone pillar (18th c. BCE)
8. Wedge-shaped symbols and pictographs made with a reed stylus on a soft writing surface, developed by the Sumerians

DOWN

1. Famous Sumerian city-state on the Euphrates River where the first plow was found and the first potter's wheel invented
2. Land between two rivers
3. Sumerian stepped pyramid that housed temples and altars to the principal god of the city
7. Goddess of love, fertility, and war, also known as Inanna
9. Symbolic foundational story about the origins and destiny of humankind

ISRAEL

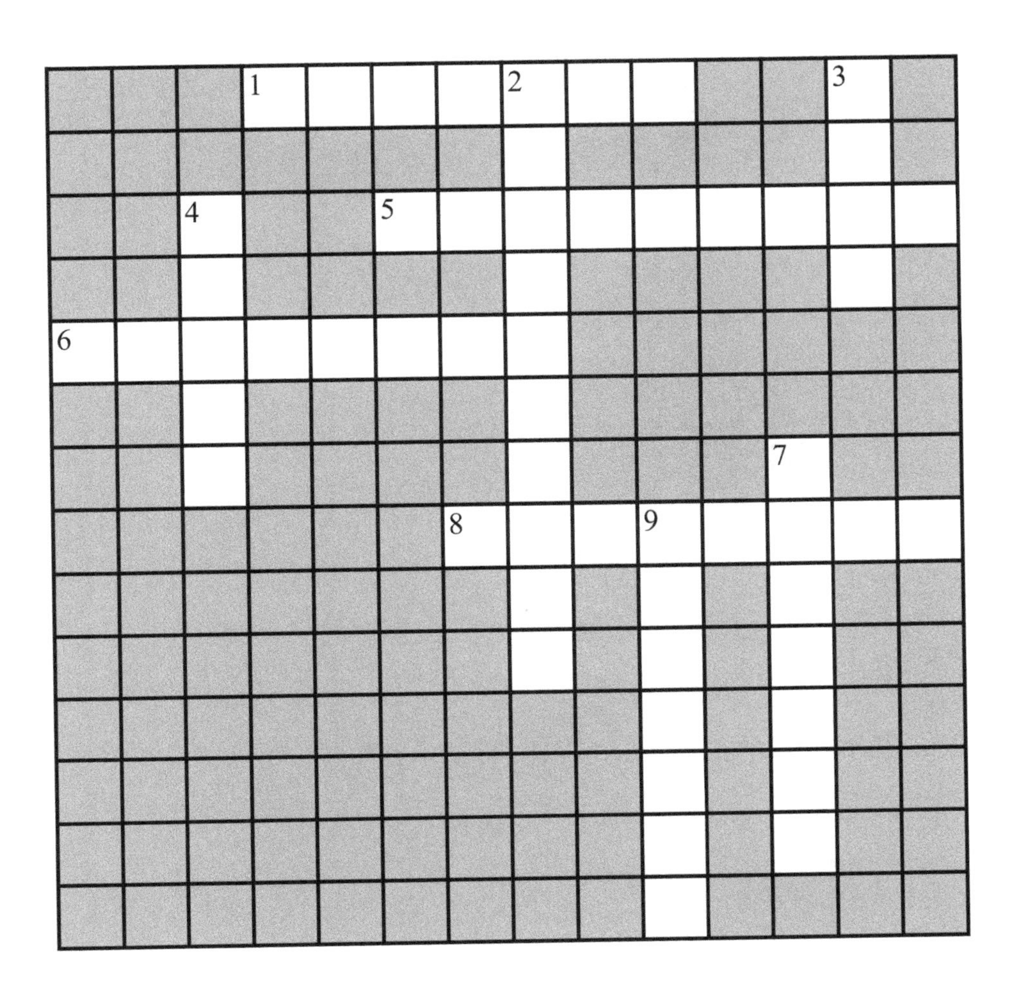

ACROSS

1. God's messenger
5. Place of worship that emerged in Babylonia after the fall of the first temple
6. Pact or binding agreement between the God of Israel and his people
8. Dispersed people

DOWN

2. Elevation of one god above other gods
3. First king of Israel
4. Second king of Israel who concentrated the cult of Yahweh in one place, Jerusalem
7. Third king of Israel who completed the construction of the very first temple in Jerusalem
9. A language family that includes Arabic, Hebrew, and the languages spoken by Babylonians and Phoenicians

Part 2: EGYPT

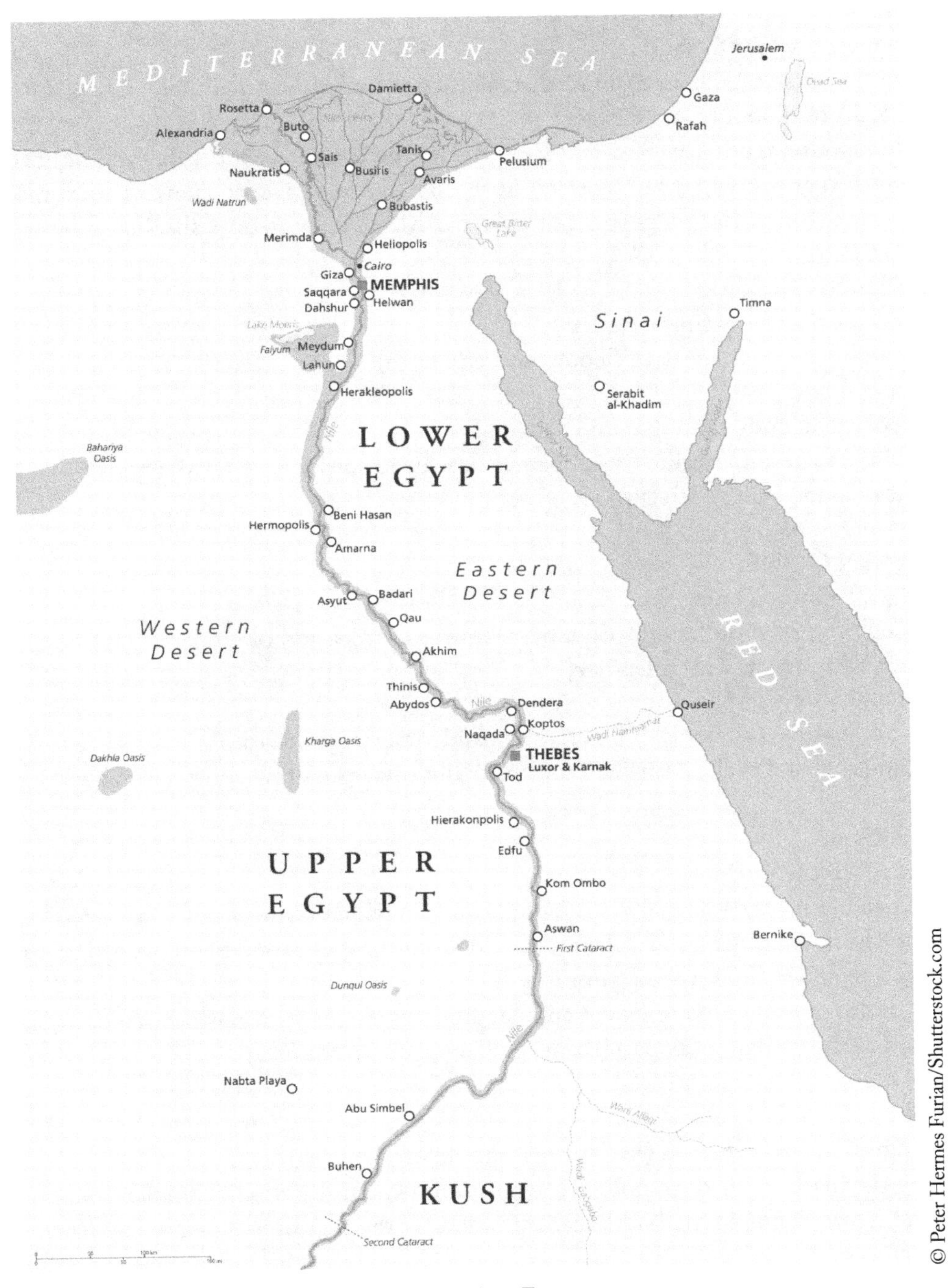

© Peter Hermes Furian/Shutterstock.com

Map of Ancient Egypt

ANCIENT EGYPT

ECOLOGY AND RELIGIOUS WORLDVIEW (DIFFERENCES)

- *EGYPT*
  - Surrounded by deserts + harborless sea coast (protected from invasion)
  - Economically self-sufficient
  - Nile River = flow predictable (flooded at right time for grain agriculture)
  - Cities less prominent (more villages)
  - Egyptian gods are orderly (Sun-god Re traverses the ocean-sky every day and fights off attacks of demonic serpents at night in Underworld so that he could be born anew the next morning)
  - King divine (son of Re: keeps order of universe, MA'AT for gods) = absolute rule
- *MESOPOTAMIA*
  - Flat, open to invasion
  - Depended on imports (more traders)
  - Tigris and Euphrates = risk of floods (unpredictable)
  - More urbanized
  - Mesopotamian gods are unfair and vengeful
  - Terrifying images of afterlife (in Gilgamesh the dead spirits slave for the gods)
  - King not divine (rules with support of nobility)

MAIN EGYPTIAN GODS

- Re (Sun-god) = Amon (rules Heaven)
- Osiris = god of vegetation (taught Egyptians how to farm) + god of Underworld [judges the dead by weighing their hearts against a feather]
- Isis, sister and wife = goddess of the earth
- Both Isis and Osiris = symbols of life, fertility, and victory over death
- Life is renewable and cyclical (in the same way Nile River goes through predictable cycles)

NILE RIVER

- Longest river in the world
- Flow predictable + after seasonal flooding leaves silt in which farmers can plant crops
- Papyrus reeds: fibers make good sails, ropes, and kind of paper exported all over Mediterranean world
- Abundant fish and animals (good game)
- Served as route of trade linking Egypt and Mediterranean Sea with northern Sudan (Nubians) and sub-Saharan Africa (Bantu people) [stone, gold, copper + turquoise]

MIGRATION

- Earlier migrations from Sahara (had turned into desert)
- Increasing population called for political organization and unification of smaller units
- Upper and Lower lands of the Nile became unified for the first time under Menes (before Mesopotamia)
- Inhabitants included various physical types from dark-skinned peoples related to sub-Saharan Africa to lighter-skinned people

BANTU PEOPLE (SUB-SAHARAN AFRICA)

- Bantu = persons or people
- Bantu-speaking people established agricultural societies south of Sahara
- Bantu languages belong to Niger-Congo family of languages (Yoruba, Wolof, Igbo, etc.)
- Pioneer-farmers + used canoe extensively in migration + iron metallurgy

TOMBS SPEAK = ADMINISTRATION AND SOCIAL STATUS

- Reflect wealth and status
  - For commoners = simple pit graves
  - For kings = pyramids built (labor tax understood as kind of religious service to ensure order)
- Egyptian history marked by tensions between monarchy and bureaucrats (reflected by tomb alignment)
  - During periods of strong monarch power, officials' tombs would be close to Pharaoh's tomb (officials served king in afterlife)
  - In times of decentralization, officials buried in home districts

TOMBS SPEAK = MATERIAL LIFE

- Depicted life of elite and also peasants
- Conventions defined status in tomb paintings
  - Obesity was a sign of wealth
  - Baldness and deformity were signs of working-class status
- Contain objects of everyday life (food, furniture) + small figurines of servants and laborers
- Contain goods + pictures that tell us whom Egyptians traded with
  - Example = gold, ivory, monkeys, baboons from Nubia

TOMBS SPEAK = WRITING SYSTEM

- Contain carved autobiographies, even texts of letters
- Hieroglyphics
  - Pictographs [picture symbols standing for words, syllables, or individual sounds]
  - Later adopted a cursive simplified script (hieratic or "priestly" script) for everyday needs

TOMBS SPEAK = WOMEN'S STATUS

- Elite women accompany husbands
- Engaged in typical domestic activities
- Depicted in yellow flesh color (men in dark red flesh color)
- Yellow flesh color indicates that elite women stayed indoors

TOMBS SPEAK = MUMMIFICATION

- Through mummification, Egyptians learned about chemistry and anatomy
- Mummification = vital organs stored in jars, body filled with spices and preserved in chemicals
- Egyptian doctors were in high demand in the ancient world
- Egyptians developed mathematics + engineering

PAPYRUS DOCUMENTS SPEAK

- Preserved in hot and dry sand
- They tell of transactions and disputes among ordinary people
- They tell that women could own, inherit, and pass their property to whomever they wished
- They contain love poetry where lovers address each other (men and women share emotions equally)

WOMEN IN EGYPT

- Enjoyed more rights than in Mesopotamia
- Marriage was usually monogamous
- Either party could dissolve marriage; divorced women retained rights over dowry
- Women could pursue trades (entertaining, nursing, and brewing beer)
- Egyptian families were often matrilineal (property descended through female line)
- Women could be priestesses and more rarely queens

THE BOOK OF THE DEAD

WHO ARE THESE EGYPTIAN GODS?

© Hein Nouwens/Shutterstock.com

© tan_tan /Shutterstock.com

Image 2 ______________________________

____

____

Image 1 ______________________________

____

____

____

Image 3 ______________________________

____

____

____

© PerseoMedusa/Shutterstock.com

Image 4 ________________________

Image 5 ________________________

ANCIENT EGYPTIAN MEDICINE

1. What were the major medical discoveries of ancient Egypt?

a.

b.

c.

d.

e.

2. What sources do historians use to reconstruct the medical history of Egypt?

a.

b.

c.

d.

ANCIENT EGYPTIAN ART

V

What did you learn today?

1.

2.

3.

4.

EGYPT

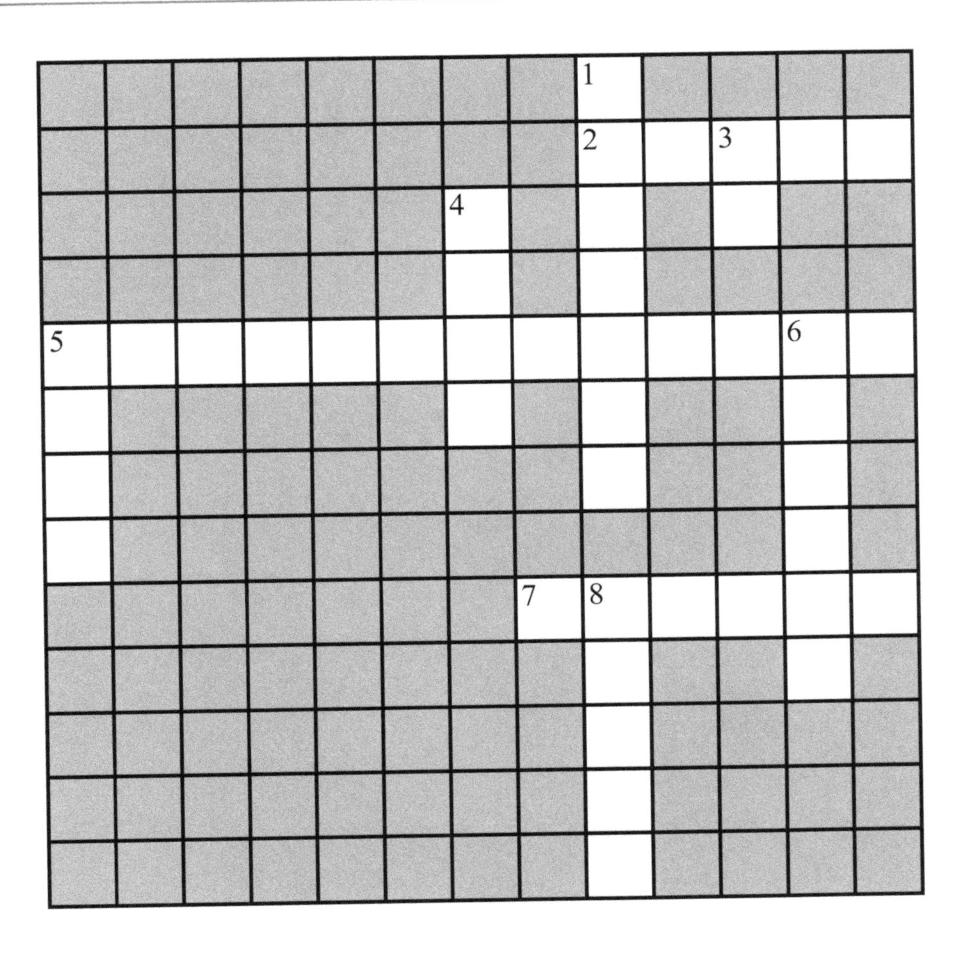

ACROSS

2. Falcon-headed god, son of Isis and Osiris, god of power and healing
5. Elaborate process for preserving the bodies of prominent people after death
7. Guardian and protector of the dead. He is associated with the embalming process and funeral rites

DOWN

1. Egyptian king considered to be god on earth and earthly guarantor of order
3. Sun-god in Egypt and chief-deity
4. Osiris's sister and wife, goddess of magic, life, and the earth
5. Egyptian term for the concept of divinely created and maintained order in the universe. The Pharaoh was the guarantor of that order
6. Prominent Egyptian god that represented the forces of nature. According to Egyptian beliefs, he taught Egyptians agriculture
8. Area south of Egypt called Kush in ancient texts. It formed its own kingdom and traded with Egypt.

Article Two

Sweet Honey in the Rocks

BY GIL STEIN, AS TOLD TO LYDIALYLE GIBSON

The history of beekeeping stretches back centuries, the director of the Oriental Institute found when a hobby turned into a scholarly pursuit.

Archaeologist Gil Stein *is director of the Oriental Institute and professor of archaeology in the Department of Near Eastern Languages and Civilizations. From 1992 through 1997, he led excavations at Hacinebi, a Mesopotamian colony in Turkey, part of the world's first-known colonial system.*

Stein is also a beekeeper. He and his wife have about a dozen hives, and their experience raising bees and collecting honey sparked his interest in the history of beekeeping, particularly in the ancient Near East. Stein spoke to the Magazine *about the insects and their Old World story.*

My wife, Liz, is the one who really got me interested—she's been a beekeeper for more than 10 years. She and I are both archaeologists, and for me it was a natural progression from intense curiosity about bees and beekeeping, and thinking how strange and wonderful this practice is, to wondering about its history. Beekeeping is pervasive in our culture and in cultures around the world. How old is it anyway? What's the archaeology of it? How did people keep bees and think about honey in the ancient world? What did it mean to them?

So I started to investigate. As I talked to people—friends who are colleagues at the Oriental Institute, who are specialists in the textual record of ancient Mesopotamia and ancient Egypt—I'd say, "Do you have any material about honey and bees and beekeeping?" And they'd say, "Yeah, we have material about honey everywhere." I'd say, "Great! Can you steer me to articles that give an overview?" And they all said no. It's just bits and pieces here and there.

Sometimes those are the most interesting problems: when something is so completely pervasive in our lives, we don't even think about it; we don't question it. Once you start looking, you realize that honey and bees and beekeeping are everywhere in the Old World—in ancient Europe and Eurasia and Africa and in the ancient Middle East. Honeybees are an Old World group of species.

Honey was considered an almost magical substance in the ancient Near East. People used it for everything: as a food and as a raw material to make alcoholic beverages like mead and honey wine. There was honey in the alcoholic beverages found in the tomb of King Midas, he of the fabled golden touch. And it's the most common ingredient in ancient medicine in Mesopotamia and Egypt. It has antimicrobial and antibiotic properties; honey will kill *Staphylococcus* and *E. coli*. It will suck the moisture out of wounds. And it's invaluable in treating burns. Ancient people also used honey as a universal sweetener, of course, because it's one of the sweetest substances in nature. They even used it for mummification. When Alexander the Great died in Babylon in 323 BC, he was preserved in honey and placed in an enormous golden sarcophagus drawn by 64 mules.

There are representations of ancient Egyptians beekeeping—tomb paintings that show people managing beehives, using techniques that are recognizable today. Once you know the artistic conventions, you can easily see it. They're applying smoke to pacify the bees and then drawing honey out of the hives. One of the clearest examples is from the Tomb of Rekhmire in ancient Thebes, which dates to the 15th century BC. That was almost three and a half thousand years ago. Beekeeping is really deep in culture.

You see honey in literature and religious texts as a common metaphor for love, for God's love for his people, and for God's law. Psalm 19 says that the Lord's ordinances are "sweeter also than honey and drippings of the honeycomb." In Exodus, God talks about delivering his people from Egypt and bringing them to "a land flowing with milk and honey."

Then there's the big question: how did beekeeping originate? The Egyptians seem to have taken it up, at an industrial scale, long before the Mesopotamians did. The earliest evidence we have of beekeeping in the Near East is from Egypt—those tomb paintings. They were also keeping bees very early on in Anatolia, which is now Turkey. Hittite laws dating to the 13th or 14th century BC contain severe punishments for thieves of bee swarms or beehives. Honey was commonly used in rituals there, and it was readily available and inexpensive; "honey bread" sold for the price of a single portion of lard or butter.

The first known mention of beekeeping in the Mesopotamian cuneiform record is centuries later. It comes from the stele of ama-re-uzur, a regional governor on the Syrian Euphrates in the middle of the eighth century BC, who claimed to have been the first among his people to capture and domesticate wild bees: "I, Samas-res-uzur, governor of the land of Suhu and Mari, I brought bees—that collect honey and which from the time of my fathers and forefathers no one had seen nor brought to the land of Suhu—down from the mountains of the Habha people and settled them in the gardens of the town of Algabbaribani."

So, did beekeeping develop independently in different parts of the ancient Near East, or did it spread from one place to another? That's one thing I'm trying to find out. I think probably there were two independent centers of invention, in Egypt and Anatolia, because there's no evidence of beekeeping in Israel for several centuries after those two places. But we don't know for sure. The evidence is spotty and scattered around.

One thing we do know is that the shapes of beehives in the ancient Near East seems to be a common technology used all over: clay cylinders laid on their sides, with a lid at one end where you would reach in and get the honey, and a little hole at the other end where the bees would fly in and out. It makes sense; that shape mimics the hollow of a tree, where many wild bees build their hives. In modern-day Egypt you can still see some of these traditional cylindrical hives, stacked up in rows.

One of the first people to pull together the information we have about ancient beekeeping was Eva Crane. Her *Archaeology of Beekeeping* [Duckworth], a wonderful book, is essentially the standard work on the subject. Since it was published in 1983, we've gotten more information. Several years ago, Israeli archaeologists working at a site called Tel Rehov, in the Jordan River Valley, excavated the remains of an Iron Age beekeeping complex, a huge apiary. At one time, there were stacks and stacks of ceramic hives. They found about 100 hives, which could have housed as many as 1.5 million bees.

For archaeologists, a huge part of the work is simply knowing what you're looking for. These ancient cylindrical beehives don't look like the box hives that most of us are used to seeing today: the Langstroth hive, which was invented in the 19th century by an American. Many people would see the remains of these ancient cylindrical hives and think, "Oh, those are roof tiles," because you see a curved shape. Or, "Those are drainpipes." I'm certain that there are many, many ancient beehives out there misidentified as drainpipes. That's why we're so lucky to have these Egyptian tomb paintings. It's undeniable proof.

I read a little bit about beekeeping almost every day. My wife and I have 11 or 12 hives, which is really small scale but still an amazing experience. Bees are such an alien species, so different from all the other domesticated animals that humans have been breeding and exploiting for millennia. We're used to cattle and pigs and chickens and goats. But enormous colonies of insects? And this stuff they create, which we steal from them? Honey and pollen, beeswax and propolis, the resin-like substance that bees use to seal the hive and keep out pests and predators. It's a very hard glue that also has incredible antibiotic properties to it, just like honey does.

And bees' social intelligence is incredible. For bees, the unit is not the individual, but the collective. A beehive has 50,000 bees, and they communicate with each other using pheromones and with what's called a "waggle dance"—used by the scout bees to tell the rest of the colony where a good source of nectar is located. The Austrian ethologist Karl von Frisch won a Nobel Prize in 1973 for figuring out the waggle dance. Bees have a division of labor and a complex social hierarchy. Virgil describes it vividly in the *Georgics*: "Some supervise the gathering of food, and work in the fields to an agreed rule: some, walled in their homes, lay the first foundations of the comb, with drops of gum taken from narcissi, and sticky glue from tree-bark, then hang the clinging wax: others lead the mature young, their nation's hope, others pack purest honey together, and swell the cells with liquid nectar: there are those whose lot is to guard the gates."

The population of a hive is not constant through the year. It peaks at about 50,000 to 60,000 in the summer, during the honey flows, and then it drops off in October and November. During the winter, a solid basketball-sized clump of bees will cluster, huddled tightly together for warmth. And they're all beating their wings constantly. Inside the hive, it can be 92 degrees in the dead of winter.

In keeping bees and doing this research, I've learned wonderful and surprising things. One of my favorites relates to the apiary at the eighth century BC site of Tel Rehov, whose excavation tells a very interesting economic story. The Jordan River Valley, where Tel Rehov is located, has a native honeybee: the Palestinian honeybee. But when entomologists looked under the scanning electron microscope at the bees they found in the residue inside the hives, those were Anatolian honeybees—a different subspecies. So the people in ancient Israel were importing honeybees all the way from Turkey, easily 1,000 kilometers away, bringing them across Syria and into the Jordan River Valley, and keeping hives of Anatolian honeybees. Because they're gentler bees and they make more honey.

So that tells you something about how economically important these insects were. People were raising them on an industrial scale and importing colonies from across the region. You can just picture some caravan transporting these bees for weeks, all the way across Syria. How could they do that? How did they keep the bees alive? But they did. If you were on the road in the ancient Near East, you might come across a bee caravan.

That's what my wife and I do too, in a way: we buy boxes of bees that get shipped to us from California. People were doing the same thing almost 3,000 years ago. That's fascinating. And what I love is, when you ask the right question, archaeologists can actually find the answer. Not every time, but often. It's amazing.

According to article 2, why is the study of beekeeping and its history so exciting for archaeologist Gil Stein? What can we learn from the history of beekeeping?

Chapter 3

People on the Move: Movement by Land (Pastoralism)

© Allocricetulus/Shutterstock.com

The Mongolian Steppe

© Mikhail Pogosov/Shutterstock.com

Two Przhevalsky Horses

© TTstudio/Shutterstock.com

Mongol Horseman in Traditional Clothing with Golden Eagles during "The Golden Eagle Festival," 2011

© Pierre-Jean Durieu/Shutterstock.com

Mongol Yurt in the Steppe

© Maxim Petrichuk/Shutterstock.com

Mongol Woman Leading a Caravan of Camels Transporting Her Dismantled Yurt

© Pierre-Jean Durieu/Shutterstock.com

Milking the Yaks in Mongolia

© Iryna Rasko/Shutterstock.com

Nomadism in the Thar Desert of India

NOMADS: NO CULTURE, REALLY?

Music in the Desert (India)

Significance: Give the names of two major collections of stories transmitted by nomads orally that became the foundation of two major religions of the world.

1. ______________________________

____

2. ______________________________

____

MOVEMENT BY LAND (PASTORALISM)

PASTORALISM

- Pastoralists are herders who are dependent chiefly on herds of domestic stock for subsistence
- Nomadic pastoralism was an alternative to settled agriculture
- Why? Severe drought killed plants, or invading armies forced farmers to flee
- Nomads colonized grassland unsuitable for early forms of agriculture but suitable for grazing
- From 4000 BCE, lifeways based on domestication of animals spread rapidly

NEGATIVE IMAGE

- Pastoralists did not write their history
- Settled people feared nomads and wrote about them negatively
- Why? Nomadic people were beyond state control
- Pastoralists-agriculturalists: separate world? NO!

CONTRIBUTIONS

- Pastoralists connected agrarian communities together through migrations, trade, and conquests
- Steppe empires controlled trade routes
- Nomadic armies exacted tributes + supplanted agrarian ruling classes [In fact, many great settled civilizations had nomadic origins]
- Farmers and nomads borrowed from each other
- Ex.: Mesopotamians invented wheel and pastoralist chariot

PASTORALISTS' GREATEST CONTRIBUTIONS

- Technological: besides invention of chariot, refinement of iron
- Spiritual: their oral tales of origins became foundation of religious mythologies
  - Hebrews (Bible)
  - Aryans of India (Vedas)

OUTER EURASIA VS. INNER EURASIA (LOCATION)

- *OUTER EURASIA*
  - Mesopotamia
  - China
  - India
- *INNER EURASIA*
  - The former Soviet Union (Russia + Ukraine + Belarus + Moldova + Baltic States + Siberia + Central Asia)
  - Sinkiang and Kansu (China's Central Asian empire)
  - Mongolia

OUTER EURASIA vs. INNER EURASIA (MAIN DIFFERENCES)

- *INNER EURASIA*
  - Ecologically poor: large steppe area, all flat

- Sparsely inhabited
- Agriculture delayed
- *OUTER EURASIA*
  - More humid and agrarian
  - Densely populated
  - Agriculture: very early

ADVANTAGES OF NOMADISM

- Nomadic societies were more egalitarian than urban settlements
- Nomadic societies were healthier, because grasslands supported less dense populations, epidemic diseases could not spread as fast
- You can always create your own tribe if you do not like the rule of a warrior-king

REGIONS WITHIN INNER EURASIA

- Tundra in north (frozen land)
- Taiga (forest)
- Steppe (grassland) = nomads
- Desert in south (agrarian oasis along rivers = centers of an important trade system between Outer Eurasia and the steppe lands of Inner Eurasia already in 2000 BCE) [famous silk roads]

MOST IMPORTANT NOMADS' ANIMALS

- Horses (domesticated in Inner Eurasia in 4000 BCE): main characteristic of Inner Asian pastoralism
- Camels (domesticated by Arabs in 4000 BCE) allowed nomads in desert to travel long distances from one oasis to the other
- Sheep + cattle + goats
- All animals produced food, clothing, shelter, traction, and transportation + life cycle determined routes

STEPPE HORSES = PRZHEVALSKY HORSES

- Small
- Sturdy + fast
- Resistant to the cold
- Can survive on grass alone
- Can dig for grass under snow
- Can run up to 60 miles a day
- Can be milked
- Provide leather and meat

SEDENTARY PEOPLE AND NOMADS

- For millennia sedentary people and nomads interacted through
  - Trade
  - Diplomacy
  - Marriage
  - Raiding
  - Warfare

- Nomads needed grains + settled people needed their horses, fur, leather, harnesses for horses
- They were even exchanges of brides to strike trade and political alliances

SEPARATE ACTIVITIES? NOT ALWAYS

- Nomads tilled small plots of land as insurance against loss of herds + peasants could abandon their poor fields for nomadic life
- Nomads had their own craftsmen. They obtained gold and other metals by trading and raiding settled civilizations + carved small figurines that they could carry (horse = favorite motif)

WHAT WAS THE BASIC UNIT OF STEPPE SOCIETY? FAMILY

- Tribe = descendants from a single ancestor
- Also open to joiners of various kind
- Alliances by
  - Oaths of brotherhood
  - Exogamic marriage
- Confederacies

INDO-EUROPEAN MIGRATIONS

- 18th-19th c. major linguistic discovery
- Speakers of Indo-European descended from common ancestors who migrated from steppe region of modern-day Ukraine and southern Russia (4500-2500 BCE)
- By 3000 BCE, Indo-Europeans acquired Sumerian knowledge of bronze metallurgy and wheel and devised hitching horses to carts

THE HITTITES (17th-16th c. BCE)

- Most influential Indo-European migrants in Anatolia and Mesopotamia
- Brought farmers and herders (farming and wool production) together in a single state and economic system
- Traded with and conquered Babylonians
- Adapted cuneiform writing to their Indo-European language
- Accepted many Mesopotamian deities in their worship

MORE ABOUT HITTITES

- Were responsible for two technological innovations
  - Light horse-drawn war chariots, which Assyrians later borrowed
  - Refinement of iron metallurgy [Mesopotamians experimented with iron metallurgy (4000-1000 BCE) but their tools were too brittle]
- Some scholars argue that the chariot and metallurgy reached China from southwest Asia thanks to the Hittites and Indo-European migrants

THE INDO-ARYANS (16TH C. BCE)

- Aryans = noble people in Sanskrit (their language)
- Migrations took centuries
- Practiced pastoral economy + limited agriculture

- Used chariots
- Early Aryans did not use writing but composed/memorized poems and songs (Vedas)

VEDAS

- Vedas = knowledge
- Vedas = collection of hymns, prayers, rituals devoted to Aryan gods (still authoritative in Hinduism)
- Vedas suggest that Aryans clashed with settled Dravidians
- Then Aryans settled down, using iron-tipped plows + adapted to climate, switching from barley to rice culture when necessary

VEDIC CASTE SYSTEM

- In late Vedic Age, the Aryans increasingly recognized four main *varnas* (castes)
- Originally their society was divided into herders and farmers led by warrior-chiefs and priests

HYKSOS (SEMITIC PEOPLE—17TH C. BCE)

- Conquered Egypt (Old Kingdom)
- Introduced to Egypt
  - Horses
  - Horse-drawn chariots
  - Bronze-tipped arrows (before Egyptians used stone-tipped weapons)
- Consequences = after defeating Hyksos, Egyptians founded new kingdom (16th-11th c. BCE) + became an imperial power, to prevent new invasions

NOMADIC POPULATIONS OF EURASIA

- Indo-Iranians
- Scythians in the 8th-2nd c. BCE
- Mongols and Turkic-speaking peoples
- Uralo-Altaic peoples (Finns and Hungarians)

SCYTHIANS

- Oldest identifiable steppe empire
- "Scythic Era" 8th-2nd c. BCE
- Western and central steppes
- Cluster of different tribes
- Spoke Iranian, Turkic, and Mongolian
- Shared intermediary lifeways between nomadism and sedentarization
- No durable capital but portable yurts

SCYTHIANS = "CENTAURS" IN GREEK MYTHS

- Fought on horses in battle, using small-size compound bows and complex horse harnesses
- Used iron metallurgy
- Controlled trade routes (in fact, thanks to them, Greek influence reached Russian lands through Greco-Scythian trade)
- Produced extraordinary portable gold sculptures of animal figures
- Settled empires used them as slaves and mercenaries to fight on their side

SCYTHIAN NOMADIC CULTURE

Write four ideas or facts you learned from watching the clip(s) today.

1.

2.

3.

4.

SCYTHIAN CULTURE

1. What is a *kurgan*?

2. What makes the digging of *kurgans* controversial?

NOMADISM

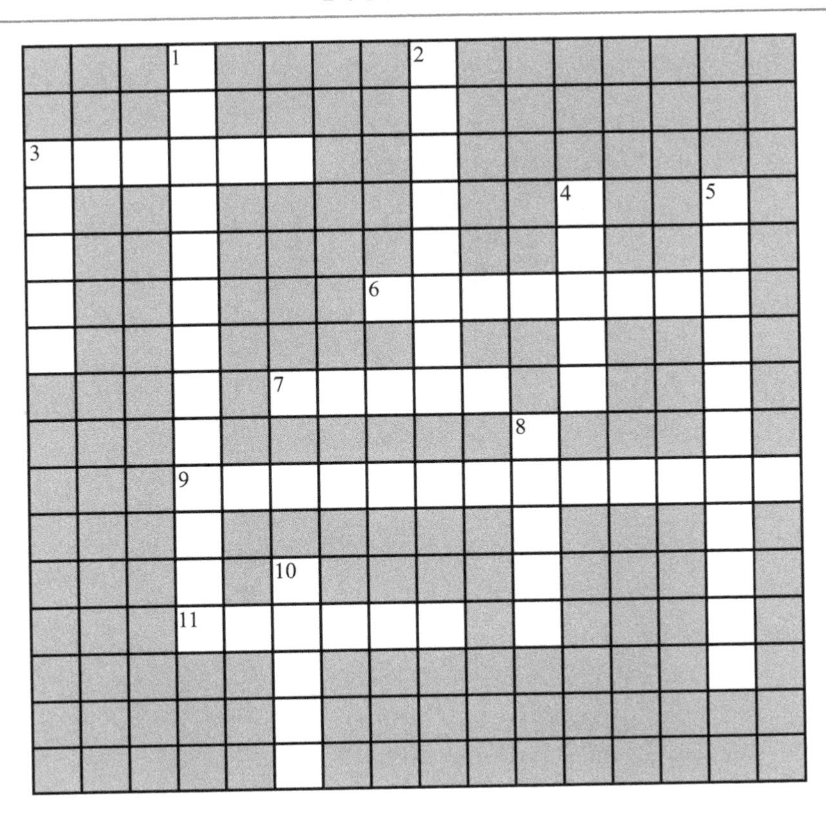

ACROSS

3. Semitic people who conquered Egypt in 17th c. BCE. As a result, Egypt became an imperial power
6. Nomads from central Anatolia who conquered Babylonians. They are well-known for two important technological inventions: light-horse drawn chariots and refinement of iron metallurgy
7. This word originally means "color" in Sanskrit (or caste)
9. A family of languages from India, Iran, and Europe that share common features (hyphenated adjective)
11. Flat grassland too arid for agriculture

DOWN

1. General term for the settled people in today's Iraq who invented the wheel (plural)
2. Oldest identifiable steppe empire
3. Most important animal for inner Asian pastoralism
4. Large association of clans that share common ancestry
5. Mode of economy involving the breeding and raising of domesticated hoofed animals or livestock. It is also a way of life dependent on moving herds of stock to new pastures throughout the year.
8. This word means "Noble" in Sanskrit (nomadic people from Central Asia)
10. This word means "Knowledge" in Sanskrit (sacred texts of the ancient Aryan invaders of India and one of the most important sacred books of Hinduism)

Article Three

China Fences in Its Nomads, and an Ancient Life Withers

BY ANDREW JACOBS

MADOI, China--If modern material comforts are the measure of success, then Gere, a 59-year-old former yak-and-sheep herder in China's western Qinghai Province, should be a happy man.

In the two years since the Chinese government forced him to sell his livestock and move into a squat concrete house here on the windswept Tibetan plateau, Gere and his family have acquired a washing machine, a refrigerator and a color television that beams Mandarin-language historical dramas into their whitewashed living room.

But Gere, who like many Tibetans uses a single name, is filled with regret. Like hundreds of thousands of pastoralists across China who have been relocated into bleak townships over the past decade, he is jobless, deeply indebted and dependent on shrinking government subsidies to buy the milk, meat and wool he once obtained from his flocks.

"We don't go hungry, but we have lost the life that our ancestors practiced for thousands of years," he said.

In what amounts to one of the most ambitious attempts made at social engineering, the Chinese government is in the final stages of a 15-year-old campaign to settle the millions of pastoralists who once roamed China's vast borderlands. By year's end, Beijing claims it will have moved the remaining 1.2 million herders into towns that provide access to schools, electricity and modern health care.

Official news accounts of the relocation rapturously depict former nomads as grateful for salvation from primitive lives. "In merely five years, herders in Qinghai who for generations roved in search of water and grass, have transcended a millennium's distance and taken enormous strides toward modernity," said a front-page article in the state-run Farmers' Daily. "The Communist Party's preferential policies for herders are like the warm spring breeze that brightens the grassland in green and reaches into the herders' hearts."

But the policies, based partly on the official view that grazing harms grasslands, are increasingly contentious. Ecologists in China and abroad say the scientific foundations of nomad resettlement are dubious. Anthropologists who have studied government-built relocation centers have documented chronic unemployment, alcoholism and the fraying of millenniums-old traditions.

Chinese economists, citing a yawning income gap between the booming eastern provinces and impoverished far west, say government planners have yet to achieve their stated goal of boosting incomes among former pastoralists.

The government has spent $3.45 billion on the most recent relocation, but most of the newly settled nomads have not fared well. Residents of cities like Beijing and Shanghai on average earn twice as much as counterparts in Tibet and Xinjiang, the western expanse that abuts Central Asia. Government figures show that the disparities have widened in recent years.

Rights advocates say the relocations are often accomplished through coercion, leaving former nomads adrift in grim, isolated hamlets. In Inner Mongolia and Tibet, protests by displaced herders occur almost weekly, prompting increasingly harsh crackdowns by security forces.

"The idea that herders destroy the grasslands is just an excuse to displace people that the Chinese government thinks have a backward way of life," said Enghebatu Togochog, the director of the Southern Mongolian Human Rights Information Center, based in New York. "They promise good jobs and nice houses, but only later do the herders discover these things are untrue."

In Xilinhot, a coal-rich swath of Inner Mongolia, resettled nomads, many illiterate, say they were deceived into signing contracts they barely understood. Among them is Tsokhochir, 63, whose wife and three daughters were among the first 100 families to move into Xin Kang village, a collection of forlorn brick houses in the shadow of two power plants and a belching steel factory that blankets them in soot.

In 2003, he says, officials forced him to sell his 20 horses and 300 sheep, and they provided him with loans to buy two milk cows imported from Australia. The family's herd has since grown to 13, but Tsokhochir says falling milk prices and costly store-bought feed means they barely break even.

An ethnic Mongolian with a deeply tanned face, Tsokhochir turns emotional as he recites grievances while his wife looks away. Ill-suited for the Mongolian steppe's punishing winters, the cows frequently catch pneumonia and their teats freeze. Frequent dust storms leave their mouths filled with grit. The government's promised feed subsidies never came.

Barred from grazing lands and lacking skills for employment in the steel mill, many Xin Kang youths have left to find work elsewhere in China. "This is not a place fit for human beings," Tsokhochir said.

Not everyone is dissatisfied. Bater, 34, a sheep merchant raised on the grasslands, lives in one of the new high-rises that line downtown Xilinhot's broad avenues. Every month or so he drives 380 miles to see customers in Beijing, on smooth highways that have replaced pitted roads. "It used to take a day to travel between my hometown and Xilinhot, and you might get stuck in a ditch," he said. "Now it takes 40 minutes." Talkative, college-educated and fluent in Mandarin, Bater criticized neighbors who he said want government subsidies but refuse to embrace the new economy, much of it centered on open-pit coal mines.

He expressed little nostalgia for the Mongolian nomad's life — foraging in droughts, sleeping in yurts and cooking on fires of dried dung. "Who needs horses now when there are cars?" he said, driving through the bustle of downtown Xilinhot. "Does America still have cowboys?"

Experts say the relocation efforts often have another goal, largely absent from official policy pronouncements: greater Communist Party control over people who have long roamed on the margins of Chinese society.

Nicholas Bequelin, the director of the East Asia division of Amnesty International, said the struggle between farmers and pastoralists is not new, but that the Chinese government had taken it to a new level. "These relocation campaigns are almost Stalinist in their range and ambition, without any regard for what the people in these communities want," he said. "In a matter of years, the government is wiping out entire indigenous cultures."

A map shows why the Communist Party has long sought to tame the pastoralists. Rangelands cover more than 40 percent of China, from Xinjiang in the far west to the expansive steppe of Inner Mongolia in the north. The lands have been the traditional home to Uighurs, Kazakhs, Manchus and an array of other ethnic minorities who have bristled at Beijing's heavy-handed rule.

For the Han Chinese majority, the people of the grasslands are a source of fascination and fear. China's most significant periods of foreign subjugation came at the hands of nomadic invaders, including Kublai Khan, whose Mongolian horseback warriors ruled China for almost a century beginning in 1271.

"These areas have always been hard to know and hard to govern by outsiders, seen as places of banditry or guerrilla warfare and home to peoples who long resisted integration," said Charlene E. Makley, an anthropologist at Reed College, in Oregon, who studies Tibetan communities in China. "But now the government feels it has the will and the resources to bring these people into the fold."

Although efforts to tame the borderlands began soon after Mao Zedong took power in 1949, they accelerated in 2000 with a modernization campaign, "Go West," that sought to rapidly transform Xinjiang and Tibetan-populated areas through enormous infrastructure investment, nomad relocations and Han Chinese migration.

The more recent "Ecological Relocation" program, started in 2003, has focused on reclaiming the region's fraying grasslands by decreasing animal grazing.

New Madoi Town, where Gere's family lives, was among the first so-called Socialist Villages constructed in the Amdo region of Qinghai Province, an overwhelmingly Tibetan area more than 13,000 feet above sea level. As resettlement gained momentum a decade ago, the government said that overgrazing was imperiling the vast watershed that nourishes the Yellow, Yangtze and the Mekong rivers, China's most important waterways. In all, the government says it has moved more than 500,000 nomads and a million animals off ecologically fragile pastureland in Qinghai Province.

Gere said he had scoffed at government claims that his 160 yaks and 400 sheep were destructive, but he had no choice other than to sell them. "Only a fool would disobey the government," he said. "Grazing our animals wasn't a problem for thousands of years yet suddenly they say it is."

Proceeds from the livestock sale and a lump sum of government compensation did not go far. Most of it went for unpaid grazing and water taxes, he said, and about $3,200 was spent building the family's new two-bedroom home.

Although policies vary from place to place, displaced herders on average pay about 30 percent of the cost of their new government-built homes, according to official figures. Most are given living subsidies, with a condition that recipients quit their nomadic ways. Gere said the family's $965 annual stipend — good for five years — was $300 less than promised. "Once the subsidies stop, I'm not sure what we will do," he said.

Many of the new homes in Madoi lack toilets or running water. Residents complain of cracked walls, leaky roofs and unfinished sidewalks. But the anger also reflects their loss of independence, the demands of a cash economy and a belief that they were displaced with false assurances that they would one day be allowed to return.

Jarmila Ptackova, an anthropologist at the Academy of Sciences in the Czech Republic who studies Tibetan resettlement communities, said the government's relocation programs had improved access to medical care and education. Some entrepreneurial Tibetans had even become wealthy, she said, but many people resent the speed and coercive aspects of the relocations. "All of these things have been decided without their participation," she said.

Such grievances play a role in social unrest, especially in Inner Mongolia and Tibet. Since 2009, more than 140 Tibetans, two dozen of them nomads, have self-immolated to protest intrusive policies, among them restrictions on religious practices and mining on environmentally delicate land. The most recent one took place on Thursday, in a city not far from Madoi.

Over the past few years, the authorities in Inner Mongolia have arrested scores of former herders, including 17 last month in Tongliao municipality who were protesting the confiscation of 10,000 acres.

This year, dozens of people from Xin Kang village, some carrying banners that read "We want to return home" and "We want survival," marched on government offices and clashed with riot police, according to the Southern Mongolian Human Rights Information Center.

Chinese scientists whose research once provided the official rationale for relocation have become increasingly critical of the government. Some, like Li Wenjun, a professor of environmental management at Peking University, have found that resettling large numbers of pastoralists into towns exacerbates poverty and worsens water scarcity.

Professor Li declined an interview request, citing political sensitivities. But in published studies, she has said that traditional grazing practices benefit the land. "We argue that a system of food production such as the nomadic pastoralism that was sustainable for centuries using very little water is the best choice," according to a recent article she wrote in the journal Land Use Policy.

Gere recently pitched his former home, a black yak-hide tent, on the side of a highway as a pit stop for Chinese tourists. "We'll serve milk tea and yak jerky," he said hopefully. Then he turned maudlin as he fiddled with a set of keys tied to his waist.

"We used to carry knives," he said. "Now we have to carry keys."

How does article 3 illustrate the perennial tensions between nomadic and settled peoples?

Article Four

Amazon Warriors Did Indeed Fight and Die Like Men

BY SIMON WORRALL, FOR NATIONAL GEOGRAPHIC

Archaeology shows that these fierce women also smoked pot, got tattoos, killed—and loved—men.

The Amazons got a bum rap in antiquity. They wore trousers. They smoked pot, covered their skin with tattoos, rode horses, and fought as hard as the guys. Legends sprang up like weeds. They cut off their breasts to fire their bows better! They mutilated or killed their boy children! Modern (mostly male) scholars continued the confabulations. The Amazons were hard-core feminists. Man haters. Delinquent mothers. Lesbians.

Drawing on a wealth of textual, artistic, and archaeological evidence, Adrienne Mayor, author of *The Amazons*, dispels these myths and takes us inside the truly wild and wonderful world of these ancient warrior women.

Talking from her home in Palo Alto, California, she explains what Johnny Depp has in common with Amazons, why the Amazon spirit is breaking out all over pop culture, and who invented trousers.

We associate the word Amazon with digital book sales these days. Tell us about the real Amazons.

The real Amazons were long believed to be purely imaginary. They were the mythical warrior women who were the archenemies of the ancient Greeks. Every Greek hero or champion, from Hercules to Theseus and Achilles, had to prove his mettle by fighting a powerful warrior queen.

We know their names: Hippolyta, Antiope, Thessalia. But they were long thought to be just travelers' tales or products of the Greek storytelling imagination. A lot of scholars still argue that. But archaeology has now proven without a doubt that there really were women fitting the description that the Greeks gave us of Amazons and warrior women.

The Greeks located them in the areas north and east of the Mediterranean on the vast steppes of Eurasia. Archaeologists have been digging up thousands of graves of people called Scythians by the Greeks. They turn out to be people whose women fought, hunted, rode horses, used bows and arrows, just like the men. (See "Masters of Gold.")

What archaeological proofs have been discovered to show that these mythical beings actually existed?

They've been excavating Scythian kurgans, which are the burial mounds of these nomadic peoples. They all had horse-centred lifestyles, ranging across vast distances from the Black Sea all the way to Mongolia. They lived in small tribes, so it makes sense that everyone in the tribe is a stakeholder. They all have to contribute to defense and to war efforts and hunting. They all have to be able to defend themselves.

The great equalizer for those peoples was the domestication of horses and the invention of horse riding, followed by the perfection of the Scythian bow, which is smaller and very powerful. If you think about it, a woman on a horse with a bow, trained since childhood, can be just as fast and as deadly as a boy or man.

Archaeologists have found skeletons buried with bows and arrows and quivers and spears and horses. At first they assumed that anyone buried with weapons in that region must have been a male warrior. But with the advent of DNA testing and other bioarchaeological scientific analysis, they've found that about one-third of all Scythian women are buried with weapons and have war injuries just like the men. The women were also buried with knives and daggers and tools. So burial with masculine-seeming grave goods is no longer taken as an indicator of a male warrior. It's overwhelming proof that there were women answering to the description of the ancient Amazons.

Why were they called Amazons?

[Laughs.] That's such a complex story that I actually devoted an entire chapter to it. It's the one thing everyone seems to think they know about Amazons: that the name has something to do with only having one breast so they could easily fire an arrow or hurl a spear. But anyone who's watched The Hunger Games, or female archers, knows that that is an absolutely physiologically ridiculous idea. Indeed, no ancient Greek artworks—and there are thousands—show a woman with one breast.

All modern scholars point out that the plural noun "Amazones" was not originally a Greek word—and has nothing to do with breasts. The notion that "Amazon" meant "without breast" was invented by the Greek historian Hellanikos in the fifth century B.C.

He tried to force a Greek meaning on the foreign loan word: a for "lack" and "mazon," which sounded a bit like the Greek word for breast. His idea was rejected by other historians of his own day, and no ancient artist bought the story. But it stuck like superglue. Two early reviews of my book even claimed I accept that false etymology. Linguists today suggest that the name derives from ancient Iranian or Caucasian roots.

You describe them as "aggressive, independent man-killers." Were Amazons also lesbians?

That is one of the ideas that have arisen in modern times. Nobody in antiquity ever suggested that. We know that the ancient Greeks and Romans were not shy about discussing homosexuality among men or women. So if that idea had been current in antiquity, someone would have mentioned it.

The one interesting artistic bit of evidence that I did find is a vase that shows a Thracian huntress giving a love gift to the Queen of the Amazons, Penthesilea. That's a strong indication that at least someone thought of the idea of a love affair between Amazons. But just because we don't have any written evidence and only that one unique vase doesn't preclude that Amazons might have had relations with each other. It's just that it has nothing to do with the ancient idea of Amazons.

The strong bond of sisterhood was a famous trait in classical art and literature about Amazons. But it was modern people who interpreted that as a sexual preference for women. That started in the 20th century. The Russian poet Marina Tsvetaeva declared that Amazons were symbolic of lesbianism in antiquity. Then others took that up. But the ancient Greeks didn't think of them as lesbians. They described them as lovers of men, actually. Man-killers—and man lovers.

You refer to the "Amazon spirit." What are its key characteristics?

I used that phrase in the dedication to a good friend of mine, Sunny Bock. She was a strong figure who believed in equality between men and women. She rode motorcycles, she rode horses, then became the first female railroad engineer. She was a risktaker who died an untimely death, probably because of her life of risk. She embodied the Amazon spirit: the assumption that women are the equals of men and that they could be just as noble and brave and heroic.

That comes through in the artworks and literature about Amazons. The Greeks were both fascinated and appalled by such independent women. They were so different from their wives and daughters. Yet there was a fascination. They were captivated by them. Pictures of Amazons on vase paintings always show them as beautiful, active, spirited, courageous, and brave.

I talked to a vase expert whose specialty is gestures on Greek vases. He has written an article about gestures begging for mercy in single combat images. Quite a few of the losers in duels are shown gesturing for mercy. But among Amazons, not so much. We have about 1,300 or so images of Amazons fighting. And only about two or three of them are gesturing for mercy. So they're shown to be extremely courageous and heroic. And I think that's the Amazon spirit.

Amazons smoked pot and drank a powerful concoction of fermented mare's milk called kumis, which they used in rituals. Put us around a campfire in ancient Scythia.

In that picture of the ancient Amazons sitting around their campfire we also have to include men. We don't have any evidence that there were whole societies with nothing but women. When we say Amazons, we mean Scythian women. In this case Scythian warrior women.

Herodotus gives us a very good picture. He says that they gathered a flower or leaves or seeds—he wasn't absolutely sure—and sat around a campfire and threw these plants onto the fire. They became intoxicated from the smoke and then would get up and dance and shout and yell with joy. It's pretty certain he was talking about hemp, because he actually does call it cannabis. He just wasn't certain whether it was the leaves or the flower or the bud. But we know they used intoxicants. Archaeologists are finding proof of this in the graves. Every Scythian man and woman was buried with a hemp-smoking kit, including a little charcoal brazier.

Herodotus also described a technique in which they would build a sauna-type arrangement of felt tents, probably in wintertime on the steppes. He describes it as like a tepee with a felt or leather canopy. They would take the hemp-smoking equipment inside the tent and get high. They've found the makings of those tents in many Scythian graves. They've also found the remains of kumis, the fermented mare's milk. I give a recipe in the book for a freezing technique they used to raise its potency. [Laughs.] Do not try this at home.

They were also very big on tattoos, weren't they?

There are a lot of tattoos—beautifully, lovingly detailed tattoos in images of Thracian and Scythian women on vase paintings. Ancient Greek historians described the tattooing practices of the culturally related tribes of Eurasia.

According to one account, Scythian women taught the Thracian women how to tattoo. The Greeks had lots of slaves from the Black Sea area, and they were all tattooed. They thought of tattoos as a sort of punishment. Who would voluntarily mark their bodies? Yet once again they had this push-pull attraction and anxiety about these foreign cultures.

We also now have archaeological evidence that Amazon-like women were tattooed. Tattoo kits been discovered in ancient Scythian burials. The frozen bodies of several heavily tattooed Scythian men and women have been recovered from graves. The famous Ice Princess is just one example—her tattoos of deer call to mind the tattoos depicted in Greek vase paintings.

Johnny Depp said, My skin is my journal, and the tattoos are the stories. I think that's a good way to think of this. They could have been initiations, they could be just for decoration, they could represent special experiences, either in reality or dreams. We don't really know. All we know is that they were heavily tattooed, mostly with real and fantastical animals and geometric designs.

A question I have been dying to ask: Who invented trousers?

The Greeks credited three different warrior women with the invention of trousers. Medea, a mythical sorceress and princess from the Caucasus region, was credited with inventing the outfit that was taken up by Scythians and Persians. The other two were Queen Semiramis, a legendary Assyrian figure, and Queen Rhodogune, which means "woman in red." The Greeks were not that far off. Trousers were invented by the people who first rode horses—and those were people from the steppes.

Leg coverings are absolutely essential if you're going to spend your life on horseback. Trousers are also the first tailored garments. They were pieced together and sewn. The Greeks wore rectangles of cloth held together with pins. They thought trousers were an abomination worn by the barbarians. But once again, they're fascinated by them. In the vase paintings the Amazons have wildly spotted and striped and checked leggings and trousers. One of the things I find most interesting is that it was not just the men who rejected trousers. Greek women didn't wear them either. Yet we find images of beautiful Amazons in trousers on women's perfume jars and jewelry boxes. I think there's something going on in Greek private life that we don't really know about yet.

There was even an Amazon island, wasn't there?

Yes. It's the only island off the southern coast of the Black Sea. It's now called Giresun Island. But it was first written about in Apollonius of Rhodes's version of the epic poem The Argonauts. As Jason and the Argonauts are sailing east on the Black Sea, they stop at what they call Island of Ares or Amazon Island. There they see the ruins of a temple and an altar, where they claim the Amazons sacrificed horses and worshipped before they went to war.

This is really interesting, because it means the Greeks were finding ruins associated with Amazons as far back as the Bronze Age. It shows how real the Amazons were to them. Recently, Turkish archaeologists found the altar and temple ruins that are mentioned in Jason and the Argonauts.

They got a bad press in the ancient world, didn't they? There were rumors that they maimed and even castrated young boys. Separate the fiction from the fact.

The idea that Amazons abandoned, maimed, or killed young boys is a fairly early story that circulated among the Greeks, because several writers assumed that Amazon societies must be women only.

That then raised the question: How do they reproduce? They came up with these stories of women agreeing to meet with neighboring tribes to reproduce. But then what did they do with the boys? So there were stories that they either maimed them so that they couldn't participate in warfare or that they actually killed them to get rid of them and only kept the girls.

The most common story was that they sent the boys back to the fathers to be raised. Many modern scholars interpreted this as proof that they abandoned their duties as mothers. They don't take care of their babies! They give them away! Blah, blah, blah.

But it turns out that it was a very common custom among nomadic people, called fosterage. Sending sons to be raised by another tribe ensures that you're going to have good relations with that tribe. It's a way of sealing treaties. It was very common in antiquity.

Philip the Great was raised by an ally of his father. It was also common in the Middle Ages in Europe. It's also a way of ensuring you don't have incest within the tribe. The fact that the Scythian and Thracian tribes probably practiced fosterage led to these stories that the Amazons gave their sons to the father's tribe. That's probably a reality. But there is no archaeological evidence that they maimed boys.

Tell us about modern-day Amazons.

Today's news from the Middle East and Syria is filled with images of Kurdish Peshmerga women fighting IS. There are movies and TV series featuring bold warrior women and even Amazons. It started with Xena: Warrior Princess, and then there were the animated films Brave and Mulan and The Hunger Games and the role of Atalanta in the Hercules film. The new Vikings TV show has all the shield maidens. And of course there are strong women in A Game of Thrones. So everyone's really aware of the idea of warrior women.

It's sort of fair to say that Amazons, both as reality and as a dream of equality, have always been with us. It's just that sometimes that fiery Amazon spirit is hidden from view or even suppressed. Right now they're blazing back into popular culture.

In article 4, how did ancient Greeks view the Amazons and why? Was it all true? Defend the Amazons!

Chapter 4

People on the Move: Movement by Sea

© Babich Alexander/Shutterstock.com

Bireme Phoenician Boat

PHYSICAL MAP OF THE MEDITERRANEAN SEA

How did ecology impact Phoenician and Hellenistic cultures?

___

___

___

___

MOVEMENT BY SEA

Besides pastoral nomadism, another alternative to farming was trade by sea

PHOENICIANS DOMINATED MEDITERRANEAN TRADE (1200 AND 800 BCE)

- Phoenicians ("purple" from dyes) occupied narrow coastal plain between Mediterranean Sea and Lebanon Mountains
- Not much land to farm, so turned to sea trade (famous for dyes + timber + iron making and importing)
- Built best ships + established commercial colonies all over Mediterranean Sea
- Spoke Semitic language and called themselves Canaanites and their land Canaan

NEW ALPHABET

- First relied on Mesopotamian cuneiform writing
- Devised new alphabet (22 symbols) representing consonants
- Writing system easier to learn than cuneiform
- More people could become literate
- Through trade Phoenician alphabet reached Greeks who added vowels. Romans later adapted Greek alphabet and passed it to heirs in Europe

MAIN COMPETITOR = EARLY GREEKS

- Mycenaeans = poor soils, so turned to sea trade
- Mountains prevented political unification; hence, organization of city-states
- Greatly influenced by Minoan culture (island of Crete)
- Power centered in palaces
- Earthquakes + herders' invasion hit walls of palaces
- Most famous king: Agamemnon (see Iliad and conflict with Troy, in today's Turkey)

ARCHAIC GREECE = MAIN CONTRIBUTIONS

- Greeks adopted Phoenician alphabet (adding vowels)
- Greeks developed several methods for governing a polis (city-state) [= city + people + surrounding countryside]
- These methods of government entered the political vocabulary of Western world:
  - Monarchy (one-man rule)
  - Oligarchy (rule by a select few)
  - Aristocracy (rule by a class of well-born families)
  - Democracy (rule by the entire body of citizens) (least efficient for Greeks)

CITY PLANNING

- In Greek city-states, voters knew leaders they elected (in Mesopotamia, governments were remote from subjects)
- Distinct feature of Greek state-cities was absence of palaces as centers of power (unlike Mycenaeans)

- City's center was an open space for general assemblies
  - Within the space: temples and administrative buildings with porches supported by rows of pillars
  - Style and techniques came originally for Egypt

OCEANIA (WESTERN HEMISPHERE)

- In Pacific Ocean, distances between islands were much longer than in Mediterranean Sea
- Colonizers of Philippines, Indonesia, Malay Peninsula came originally from indigenous people of Taiwan
- Languages belonged to the Austronesian family

LAPITA SYSTEM (AROUND 1600 BCE)

- Definition: system of kinship-based exchanges among the inhabitants of thousands of islands
- Lapita system = one of the world's largest trading spheres, stretching thousands of miles
- It was based on trade of potteries and obsidian (a hard, dark, glasslike volcanic rock formed by the rapid solidification of lava without crystallization used for tools and weapons in the absence of metals)

HOW WAS TRAVEL MADE POSSIBLE?

- Technological advances in long-distance navigation = sail- and paddle-drive oceangoing canoes
- Creation of extensive, orally transmitted navigational information (still retained by Polynesian elders) = mental maps of islands and currents
- Cultivation of root crops (yams and taros) easily stored for travel
- By 16th c., every island had been settled. When Europeans arrived, they brought diseases (native population reduced to almost 80 percent)

1. The Phoenicians or Canaanites living on the coast of modern Lebanon and Syria notably developed which of the following?

   a. Steel b. Vowels c. Pictographic alphabet d. Consonantal alphabet

2. Retrace the genealogy of our alphabet.

   Greece • • 1

   Phoenicia • • 2

   Rome • • 3

3. The word *Phoenician* comes from a Greek word which means

   a. Green b. Yellow c. Purple

4. Why did the Greeks call the Canaanites *Phoenicians*?

5. Which of these governments did the Greeks consider the most inefficient?

   a. Oligarchy

   b. Democracy

   c. Monarchy

   d. Aristocracy

POLYNESIAN CULTURE IN THE PACIFIC OCEAN: WHAT SOURCES DO HISTORIANS USE TO RECONSTRUCT THE PAST?

1.

2.

3.

4.

Chapter 5
Early Societies in the Americas

© Guzel Studio/Shutterstock.com

The Mayan Temple of Kukulkan (Chichen Itza, Mexico)

EARLY SOCIETIES IN THE AMERICAS

ARRIVAL OF AMERINDS (= AMERICAN INDIANS)

- Human life originated in eastern Africa
- North and South Americas were last continents to be populated by humans
- Homo sapiens reached Western Hemisphere during last ice age (when glaciers locked up much of the earth's water) [12,000-10,000 BCE]

ORIGINS OF AMERINDS

- On basis of linguistic and biological evidence, Amerinds are related to Mongolian peoples of Asia
- Type O blood carrying a specific antigen found only in Mongolian peoples
- Native Americans do not accept this reading
- Emergence or creation myths tell of their beginnings in Western Hemisphere

EASTERN vs WESTERN HEMISPHERES (COMMONALITIES)

- Move from foraging to agriculture and domestication of animals
- Move toward urbanization
- Specialized labor
- Hierarchical social order
- Presence of long-distance trade
- Organized religion
- Development of monumental architecture for religious and political purposes + traditions of writing

MAIN DIFFERENCES

- *WESTERN HEMISPHERE*
  - Cultural transmission and economic exchange travelled much slower (different climate zones north to south, east to west)
  - Most large animals could not be domesticated (no horses, no oxen)
  - No need for wheel
  - Disease environment benign
  - Little writing (did not travel well)
- *EASTERN HEMISPHERE*
  - Cultural transmission and economic exchange travelled fast across zones of similar climate (no need to adapt)
  - Most large animals could be domesticated (horses, oxen)
  - Wheel
  - Many human diseases originated in livestock (measles, smallpox, influenza)
  - Writing (traveled faster: Mesopotamia, Phoenicians, Greeks, Romans, Europeans)

AMERICAS AND AFRICA (except Egypt) HAVE A LOT IN COMMON

- Agricultural transformation occurred roughly at the same time

- Rough environmental conditions (different climate zones + few rivers + dense forests) favored
- Low population density
- Delayed agricultural transformation
- Slowed transition to city-states

THE OLMECS OR THE RUBBER PEOPLE IN MESOAMERICA (1200-900 BCE)

- Culture influenced Mayans
- Cultivated maize and corn + used obsidian tools to clear the forest
- Were famous for drainage constructions to divert floods
- Created calendar to keep track of seasons

RELIGION

- Their main urban centers were more ceremonial centers than true cities
- Priests, administrators, and craftspeople inhabited center + centers surrounded by villages (farmers served as laborers to build temples)
- Created a system of writing (calendrical inscriptions) = oldest writing system in Americas (similar to Egyptian hieroglyphics)
- Practiced human sacrifices and invented team ball game

MAYANS (HEIRS OF THE OLMECS)

- Southern Mexico, Guatemala, Belize, Honduras, and El Salvador
- Mayans were not isolated cultures but nobility traded jade, obsidian, and cacao bean that makes chocolate
- Cacao beans were so precious that Mayans used them as money
- Borrowed from Olmecs (astronomy, mathematics)

MAYANS ADDED OWN TOUCH

- Created numerical system using zero as a number (unknown to Romans) = easier to manipulate large numbers
- Calculated length of solar year
- Understood movement of heavenly bodies (could predict eclipses of the sun and moon)

WRITING

- On stone pillars and long strips of tree bark paper or deerskin
- Ideograms (like Chinese characters) and symbols for syllables
- Script deciphered in 1960s
- Wrote works of history and poetry + kept genealogical, administrative, and astronomical records
- Invading Spaniards burned most of their books to undermine native religious belief

RELIGION AND AGRICULTURE

- Popol Vuh = Mayan creation myth = gods had created humans out of maize (flesh) and water (blood)
- Gods kept world going and maintained agricultural seasons in exchange for sacrifices and bloodletting rituals

- Shedding blood would prompt the gods to send rain to water the crops
- Bloodletting rituals = war captives but also voluntary shedding of royal blood (male or female)

RELIGION AND ARCHITECTURE

- Myth: earth is flat with thirteen layers of heaven above it and nine levels of underworld. A sacred tree rose through the different sections
- Mayan pyramids represented this cosmology: top reached the heavens and portals were doorways to the Underworld

MAYAN TEAM BALL GAMES

- Unique in ancient world
- Stadiums (with elevated platforms for better view)
- Heavy ball made of solid baked rubber
- Players wore protective gears around their hips and helmets
- The goal was to propel ball through a ring without using hands
- All Mayan ceremonial centers had stone paved courts for the game
- For fun (bets, celebration of a treaty) or game pitted high-ranking captives against each other (losers became sacrificial victims)

HOHOKAM IN SONORAN DESERT

- Hohokam of Sonoran Desert in Arizona probably migrated from Mexico
- Taste for ball courts attest to connections with Mesoamerica (village teams competed with one another)
- Pioneer farmers (they grew maize, corn, squash, cotton)
- Constructed canals for irrigation
- We, in Phoenix, inherited their canal irrigations

MAYAN CIVILIZATION

1. Where are the Mayans located?

2. Who were the Olmecs? Why are they significant?

3. What were the major achievements of Mayan civilization?

4. What delayed the deciphering of the Mayan writing system?

5. How did the deciphering of the Mayan writing system change our perception of Mayan culture?

THE WESTERN HEMISPHERE

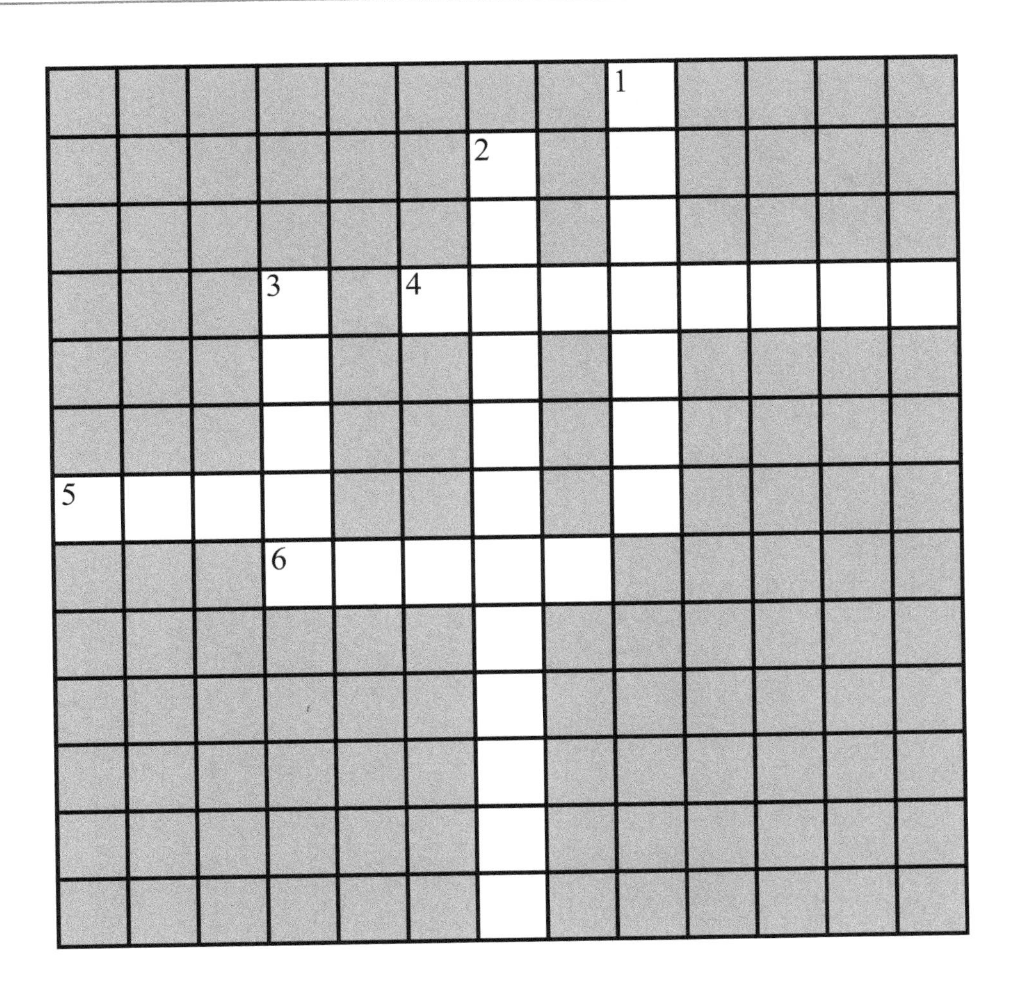

ACROSS

4. Famous Mayan creation myth
5. People of Southern Mexico who invented the zero as a number.
6. "Rubber People" of Mesoamerica whose culture greatly influenced the Mayans

DOWN

1. People in Arizona who built canal irrigations still used today
2. Famous Mayan fertility ritual that connected the temporal and spiritual worlds
3. Famous beans used as currency among Mayans

Chapter 6

Introduction to the Religious Foundations of Empires

Giant Buddha Statue in Tian Tan (Hong Kong, China)

© Renata Sedmakova/Shutterstock.com

Statue of Plato in Front of the National Academy of Athens by the Italian Sculptor Piccarelli (19th c.)

THE AXIS AGE

MAIN PATTERNS UNTIL 6TH C. BCE

- Pattern of urbanization started in Mesopotamia + developed independently in Western Hemisphere
- Cities brought together people from different tribal cultures
- Agrarian surpluses allowed people to engage in different occupations (carpentry, blacksmithing, record keeping)
- Emergence of classes (peasants, craftsmen, noblemen, priests, slaves)
- In a parallel fashion, people assigned special powers and tasks to their former agrarian gods, now gods of the city

FORAGING AND URBAN COMMUNITIES (DIFFERENCES)

- *FORAGING COMMUNITIES*
  - Shared a collective identity (same tribe, same stories, same rituals)
  - Everyone shares the tasks of hunting and gathering
  - More gender and social equality
  - The dead took their place among sacred ancestors. They continued to dwell with the living in a single community
- *URBAN COMMUNITIES*
  - Cities brought together people out of different tribal cultures. People brought different stories, rituals, and family identities
  - Labor became much more specialized
  - Rise of individualism + gender and social inequality
  - What happens to the self after it dies? Is life still worth living if it is filled with social injustice?

AXIS (AXIAL) AGE [COINED BY KARL JASPERS] = CA. 6TH C. BCE

Produced most influential philosophies and religions of the world

- Early Greek philosophers (Plato)
- The Upanishads in ancient India (basis for Indian philosophy)
- The Buddha in ancient India
- Mahavira (Jainism) in ancient India
- The Prophets of Israel + time of fall of first temple + Babylonian exile
- Zarathustra/Zoroaster in Iran (Zoroastrianism)
- Confucius and Laozi in China

MAIN CHARACTERISTICS

- These religions/philosophies spread beyond place of origin and moved along axis (southern Eurasia-northern Eurasia)
- Huge following + disciples left writings which became references for further reading and interpretation
- Raised questions about human nature + place in universe
- Arose at time of political fragmentation
- Devised new solutions for preserving social and political order

MOVE AWAY FROM POLYTHEISM

- Moved away from mythological forms of thinking centered around unpredictable multiple gods to understanding universe without primary reference to them
- Moved away from ancient belief in many gods to search for a single abstract reality as source of everything

MOVE FROM DEPENDENCY ON PRIESTLY CLASS TO INDIVIDUAL THINKING

- Moved away from sacrifices to the gods to keep them happy
- Some rejected religious rituals and tried new techniques for altering consciousness (meditation, asceticism)
- People of any social class (not just priests who were in charge of temples and sacrifices) could potentially take this path

RELIGION (DEFINITION)

- Bruce Lincoln (American historian of religion)
- Religion is:
  - a discourse (based on foundational myths)
  - a set of practices (rituals, rules of behavior)
  - a community (members construct their identity with reference to a religious discourse and its practices)
  - an institution (regulates religious discourse, practices, and community)

GREAT RELIGIONS OF THE WORLD

- **Myths of Nature** [Earliest religions]
- **Myth of Liberation** [Religions originating in South Asia (Hinduism, Buddhism, and Jainism)]
- **Myth of Harmony** [Religions originating in China, Daoism and Confucianism]
- **Myth of History** [Religions of Middle East (Judaism, Christianity, Islam)]

MYTHS OF NATURE

- Stories about forces of nature that govern human destiny. Forces of nature can be personified (gods, spirits, animals, sacred ancestors)
- Hunter-gatherer and agrarian stories emphasized fertility of the earth + need for the ritual renewal of life in harmony with seasons
- Shaman = spiritual leader, making trance journeys to the spirit world to restore harmony with nature

RELIGIONS OF SOUTH ASIA AND MYTH OF LIBERATION

- Life is seen as suffering
- Why? Life always ends in old age, sickness, and death
- Humans are caught in an endless cycle of suffering, death, and rebirth (*samsara*)
- The goal of religion is to free oneself from *samsara*
- Once freed humans will experience ultimate reality

CHINESE RELIGIONS AND MYTH OF HARMONY

- Tao (Dao) = mysterious source and ordering principle of the universe
- All of creation works via the opposites of *yin* and *yang*, of dark and light, of earth and heaven, of female and male
- The ideal of life is balance and harmony
- Problem of existence is disharmony
- Daoism and Confucianism offer different means to reestablish harmony

RELIGIONS OF MIDDLE EAST AND MYTH OF HISTORY

- God is the creator of all things
- God acts in time and leads his people through time toward a final fulfillment
- The story begins with an initial harmony
- Then this harmony is disrupted by sin
- Judaism, Christianity, and Islam trace themselves back to Patriarch Abraham
- In all three, the goal is to restore harmony with the will of God so that death can be overcome

GREEK PHILOSOPHERS

- Applied reason to the physical world
- Rejected mythical explanations for natural phenomena
- Capricious gods did not manipulate nature
- Principles of order under seeming chaos of nature (can be investigated through reason)
- Greatly influenced Abrahamic religions

CLICKER NOTES

Write a paragraph about the significance of the Axis Age.

1. TIME

2. PLACE

3. DEFINITION

4. SIGNIFICANCE

Article Five

Thumbing Your Nose at Zeus

BY SIMON WORRALL, FOR NATIONAL GEOGRAPHIC

The same god was worshipped differently in each city. Some Greeks worshipped none at all.

By Carroll, Christopher

In ancient Athens there was a dinner society called the Bad Luck Club. Determined to mock the gods and the laws of the city, its members were said to have scheduled their private meals on ill-omened days, when feasting was forbidden. Having thus piqued the gods, all of them died miserable deaths, except for one lone survivor, whose life, we are told in an account from the third century A.D., was more a punishment than death would have been anyway. Stories like this one, written not by atheists but by pious contemporaries who often regarded them with suspicion, make up much of the evidence that attests to atheism in ancient Greece.

Few texts by ancient disbelievers have survived. And yet, as Tim Whitmarsh argues in "Battling the Gods: Atheism in the Ancient World," "atheism has a tradition that is comparable in its antiquity to Judaism (and considerably older than Christianity or Islam)." Mr. Whitmarsh, a classicist at the University of Cambridge, has undertaken to tell the story of Greek atheism over a thousand-year period. Drawing on the close reading of many texts, he suggests that the tradition of Greek disbelief is more considerable than previously imagined, that Greek atheists were "airbrushed out of ancient history, or their significance minimized." His book, which ranges from the dark ages of Greece to the imposition of Christianity as the sole legal religion of the Roman Empire in the fifth century A.D., is a remarkable survey of the ways in which the Greeks questioned and rejected notions of the divine, as impressive for its breadth and erudition as for the concision, clarity and ease with which it conveys a sometimes forbiddingly complex story.

For much of Greek antiquity, Mr. Whitmarsh writes, atheism wasn't treated as a heresy but was "seen rather as one of the many possible stances one could take on the question of the gods (albeit an extreme one)." This was in large part owing to the pluralistic nature of ancient Greek society. Classical Greece was a world of squabbling city-states spread around the Mediterranean, as Plato wrote, like "frogs around a pond." Just as they had no one ruler, the Greeks had no one sacred text--no equivalent of the Tanakh, the Bible or the Quran. The pantheon, established mainly by the epics of Homer and Hesiod, was roughly the same for all Greeks, and yet always different. "In Greek polytheism," Mr. Whitmarsh explains, "religious ritual is always localized: you pray not to Athena as an abstract deity but as her specific manifestation in your local sanctuary." And in each city, the same god would be worshiped differently. Consider the example of Artemis, who, Mr. Whitmarsh writes, at Brauron near Athens presided over a ritual involving young girls of marriageable age dressing as bears; who near Ephesus on the Anatolian coast occupied the largest temple in the region and was depicted in the guise of a pre-Greek deity with a profusion of what have been variously interpreted as breasts, eggs, or even bull's testicles; and who at Patrae was worshipped, as Artemis Laphria, with a huge fire onto which were thrown wild animals of all kinds.

Because there was no universally imposed orthodoxy--no canonical idea of what exactly the gods were like or how one was meant to worship them--it wasn't considered blasphemous to question their nature. In one of the most impressive chapters of his book, Mr. Whitmarsh describes how the many pre-Socratic philosophers, beginning in the sixth century B.C., attempted to explain the world around them not through Homeric or Hesiodic myth but by the observation of physical reality. Thus for the philosopher Anaximenes the stars were not supernatural but "pieces of Earth that had been borne aloft on evaporated moisture and had subsequently caught fire." Yet these philosophers were not radical atheists--they didn't deny the existence of divinity. They sought instead to redefine it, often as a sort of omnipotent, animating principle.

By the 420s, Mr. Whitmarsh notes, tragedians and comedians were writing plays exploring not just the gods' nature but whether they existed at all. A number of these plays contain arguments against theism that are still used today. The "Bellerophon," for instance, a tragedy by Euripides that survives only in fragmentary form, contains an early version of what philosophers now call the Problem of Evil:

Someone says that there really are gods in heaven?
There are not, there are not--if you are willing
Not to subscribe foolishly to the antiquated account.
Consider it for yourselves; do not use my words
As a guide for your opinion. I reckon that tyrants
Kill very many people and deprive them of their property
And break their oaths to sack cities;
And despite this they prosper more
Than those who live piously in peace every day.

Though these plays, as Mr. Whitmarsh notes, ultimately affirmed prevailing beliefs, they were presented in front of audiences of thousands of citizens, providing a means of considering ideas about the gods introduced by pre-Socratics and sophists.

This is not to say that it was entirely a good idea to go about denying the existence of the gods. Though polytheism was generally tolerant of disbelief, there were periods of repression, the first of which came in imperial Athens during the Peloponnesian war in the fifth century, as Athens's fortunes began to wane and the citizenry grew skittish about offending the gods. A number of philosophers, Socrates most prominent among them, were tried and convicted on highly politicized charges of atheism and impiety, and it was in this period that the Greek word atheos, which had first meant something more like godforsaken, took on a second meaning of "atheist."

This suppression, though it had lasting effects, did not snuff out ancient disbelief. Mr. Whitmarsh's account continues through the conquests of Alexander the Great, the Hellenistic age and Greece's absorption into the rising Roman Empire, finally ending in the Christian era. It was only then, Mr. Whitmarsh writes, that "atheism began to be constructed in systematically antithetical terms, as the inverse of proper religion, a threat to the very foundations of human civilization."

Mr. Whitmarsh's book is, on the whole, a delight to read, though his attempts to show that atheism was more prominent than we may have thought can be excessively speculative. One such example is the story of Salmoneus, who was said to have been killed by the gods after claiming to be Zeus and dragging dried hides with bronze kettles behind his chariot in an attempt to mimic the sound of thunder. Mr. Whitmarsh sees in the story "a meditation on the metaphysical implications of a culture that was beginning to manufacture divinity in the human realm, through sculpture, painting, and theater." Others may not see this at all. Still, even if a reader were to disagree with every such suggestion, Mr. Whitmarsh's book would be very much worth attention, a sophisticated and nuanced account of a fascinating and too often overlooked world.

According to article 5, how did ancient Greeks relate to their gods?

Chapter 7

Hinduism

Statue of Vishnu on Hindu Temple Wall

India Today

AFGHANISTAN
Harappa
Srinagar
CHINA
PAKISTAN
Amritsar
Indus River
NEW DELHI
Mohenjo Daro
NEPAL
Jaipur
Agra
BHUTAN
Jaisalmer
Kanpur
Ganges
Imphal
Udaipur
INDIA
Varanasi
Kandla
BANGLADESH
Ahmedabad
Bhopal
Kolkata
BURMA
(MYANMAR)
Nagpur
Arabian
Sea
Mumbai
Hyderabad
Visakhapatnam
Panaji
Bay of
Bengal
Mormugao
Bangalore
Chennai
Andaman
Islands
Kozhikode
Pondicherry
Port Blair
Kochi
Madurai
Thiruvananthapuram
Nicobar Islands
Kovalam
SRI LANKA

WHAT IS MISSING?

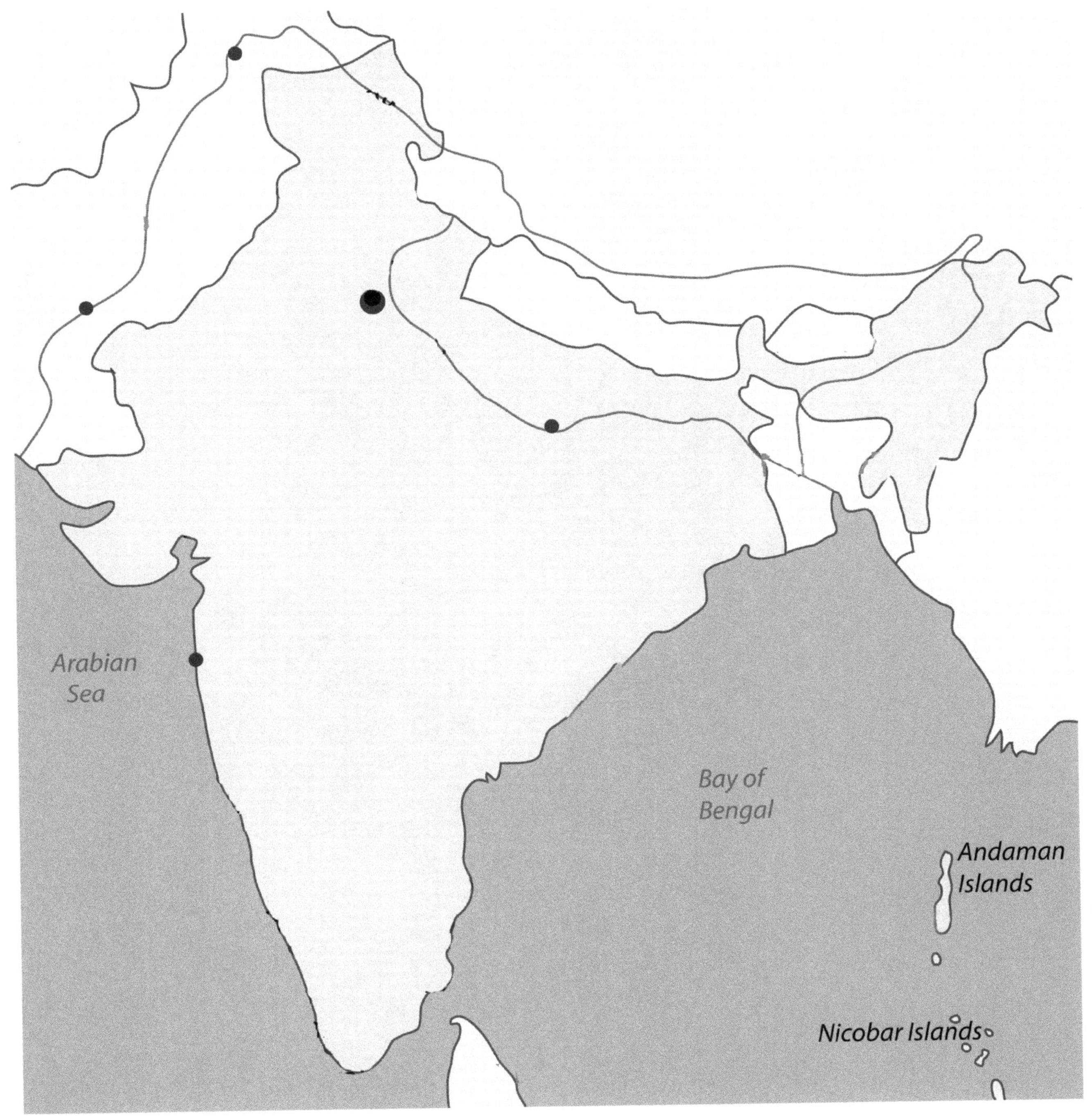

Instructions: Please locate Pakistan, India, New Delhi, Varanasi, Mumbai, Harappa, Mohenjo Daro, the Ganges and Indus Rivers on the map.

THE CASTE SYSTEM (VARNA)

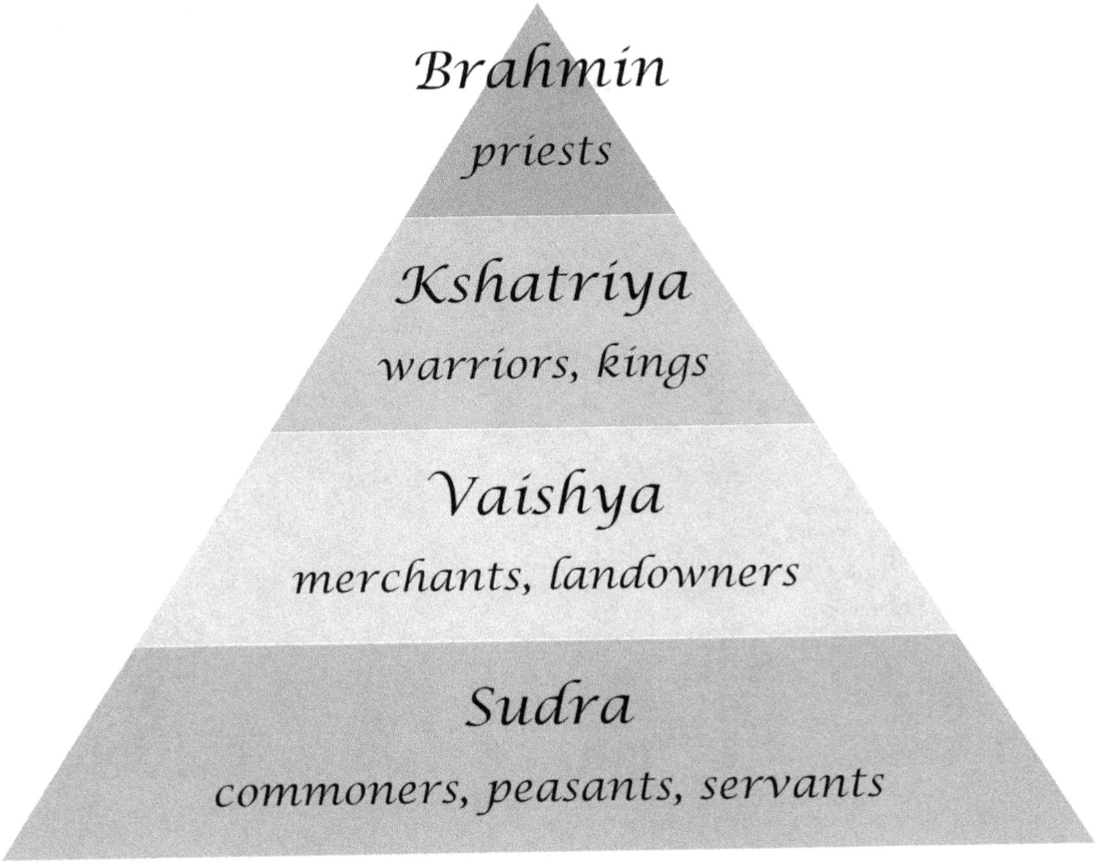

Brahmin
priests

Kshatriya
warriors, kings

Vaishya
merchants, landowners

Sudra
commoners, peasants, servants

What was the name of Arjuna's caste? ______________________________

Child Dressed as Lord Krishna in Bangladesh (2015)

Who is Lord Krishna? He is the ______________________________ (alternate form) of the God ______________________________, also known as the Preserver.

HINDUISM

HINDUISM

- Third largest religion today after Christianity and Islam
- From Sindhu (Indus River) in Sanskrit
- Amalgam of spiritual traditions originating in South Asia
- Unites
  - Worship of many gods
  - Belief in a single divine reality

CONTRARY TO CHRISTIANITY AND ISLAM, HINDUISM DOES NOT HAVE

- an identifiable founder
- a single canonical text accepted by all followers
- a religious elite or single priestly group
- a strong organizational structure to spread influence

HARAPPA CIVILIZATION

- When? Before Axis Age
- Where? In the Indus Valley (in modern-day Pakistan)
- Significance:
  - Important agrarian center (= Mesopotamia and Egypt)
  - Religious symbols found at Harappa still appear in contemporary Indian culture

ARYAN INVASION?

- When? About 1500 BCE (before Axis Age)
- Nomads (herders) from Central Asia
- Forced Dravidians (indigenous population) into southern India
- Aryans = nobles in Sanskrit
- Language: Sanskrit
- Sacred writings = the Vedas

THE FOUR VEDAS

- Vedas = knowledge
- Earliest sacred books of Hinduism
- ca. 1500-400 BCE
- Liturgical handbook of early Aryan priests (Brahmins)
- First existed orally

RIG VEDA

- = "hymn knowledge"
- Oldest of the four Vedas
- Over 1,000 hymns addressed to a single or to two or more Aryan deities called devas (deus, in Latin).
- Probably for the elite

EARLY INDIC RELIGION IN VEDAS

- No permanent temples
- Worship centered on fire sacrifices under the open sky

BRAHMINS (PRIESTS)

- Led sacrifices to Aryan deities
- Chanted the Vedic hymns
- Brahman
  - = holy word, sacred knowledge, incantation, mystic utterance
  - = stanzas of the Rig Veda
- Brahmins = the ones who recited *brahman*
- To become a Brahmin initiation needed

SACRIFICES TO THE GODS

- Gods had to be kept happy
- Priests
  - Chanted Vedic hymns
  - Offered grain, animal flesh, and melted butter into a fire
- Each priest had a special function
  - Altar builder
  - Libation pourer and invoker of the gods
  - Fire kindler

COSMOGONIC MYTH

- Primordial being Purusha sacrificed
  - His mouth became the priests
  - His arms the warriors
  - His thighs the producers
  - His feet the workers
- First allusion to the fourfold class (*varna*) system (caste system)

ARYAN INFLUENCE

- The Aryan class system came to dominate all of India
- Became basis of the later caste (*varna*) and subcaste (*jati*) system
- Each *jati* has proper occupation, duties, and rituals

VEDIC DEITIES

- Major deities of the Vedic world = all male [Dravidian deities were mainly female]
- Connected to
  - sacrifice
  - martial conquest
  - mystical experience
  - maintenance of moral order

- Two kinds of deities:
  - Deities of earth and skies
  - Liturgical deities (= deified elements of ritual)

DEITIES OF EARTH AND SKIES

- INDRA = a warrior atmospheric god (every night fights demon that encases the universe)
- VARUNA = sky god and sustainer of the cosmic order + administers justice
- RUDRA-SHIVA
  - = the dread mountain god
  - = destroyer or healer?
  - = worshippers asked him to be auspicious (*shiva*)
  - = significance: early form of Hinduism's great god Shiva, the Destroyer and Reviver

LITURGICAL DEITY = AGNI

- = fire god + intermediary between gods and worshipers
- Cleansed
- Removed sin and guilt
- Drove away demons
- Protected homes

MAIN CHARACTERISTICS OF VEDIC TRADITION

- Polytheistic system
- Did not elevate one deity above the others
- During prayers, they spoke of each of their divinities as being supreme (henotheism)
- Henotheism = temporary elevation of one of many gods to the highest rank that can be accorded, verbally or ritualistically

AXIS AGE (6TH BCE)

- New thinkers sought a single divine reality that might be the source of everything
  - Some came to reject religious ritual
  - Others abandoned social and family life to live alone in the forests
  - Others experimented new meditation techniques to seek the divine reality
  - People of any social class could experiment these techniques, not just priests
- Buddhism and Jainism came to be

THE UPANISHADS DURING THE AXIS AGE (6TH C. BCE)

- Four Vedas end with later works, the Upanishads
- Upanishad = to sit nearby
- Upanishads = religious dialogues
- Literature for the elite
- Main theme: relationship between the individual self and Brahman (underlying support of all)

SIGNIFICANCE OF UPANISHADS

- Emphasis placed on inner experience

- Goals
  - To seek intuitive knowledge of that which is the source of all reality
  - To reach spiritual immortality
- In Vedic religion, emphasis placed on outward ritual performances
  - Goal = long life, wealth, good health

NEW EMPHASIS ON THE SOUL

- Soul = one's inner Self (*Atman*)
- *ATMAN* is not *jiva*
- *Jiva* = observable, empirical self with a small "s"

SALVATION THROUGH MEDITATION

- Upanishads distinguished between
  - *Atman*, the nonmaterial inner Self, and
  - Matter, that is, the natural world
- Matter is of an inferior order
- To be content with the natural world can only result in suffering
- Salvation is best attained by breaking away from the natural world
- Meditation will help free soul from body

NEW EXPERIMENTS IN MYSTICAL TECHNIQUES

- Could be done by people of all classes, not just the priests
- Did this change occur because some castes rebelled against the rigid Brahmin control of all life?

IMPORTANT CONCEPTS OF UPANISHADS

- Brahman
- Atman
- Maya
- Karma
- Samsara
- Moksha

BRAHMAN

- = the holy power of prayer in Vedic literature
- = the ultimate reality of the world in the Upanishads
- Usually Brahman pictured as impersonal reality
- How can Brahman be known? By meditation through Atman

ATMAN

- Brahman became subdivided into myriad individual Atmans
- *Atman* = the innermost, unseen, transcendental Self of a person
- It is different from *jiva*, the observable, empirical self with a small "s"

MAYA

- *Maya* = illusion = the world viewed inadequately
- In truth, *Atman*, the eternal soul, is one with *Brahman*
- The basic human problem is ignorance
- Because of our ignorance, we believe that our true nature is the self (*jiva*) that does act + we do not know the true, spiritual reality of *Atman/Brahman*

KARMA

- To experience the joining of *Atman* and *Brahman*, the devotees have to free themselves from the self that acts (that is, the Karmic self or *jiva*)
- *Karma* = action and consequences of an action in Sanskrit

SAMSARA

- = wheel of birth, death, and rebirth
- Because of our ignorance, the *Atman*, the true Self is trapped in the cycle of rebirth

MOKSHA

- The ultimate goal is NOT creation of good lives by good deeds
- The goal is to escape from *Samsara* (cycle of rebirth)
- How? Seek true knowledge.
- True knowledge involves an experience of Oneness
- It is done through self-discipline and meditation

THE UPANISHAD REVOLUTION

- Interiorization of religiosity
- Individual's struggle to free the inner Self (*Atman*) from matter replaced Vedic communal sacrificial rites

SUMMARY

- Before 6th c. BCE = Vedas = sacrifices to Gods + polytheism/henotheism + caste inequality
- 6th c. BCE = Axis Age = time of the Upanishads (new sacred writings) = search for single reality through meditation = anyone can do it!

CLASSICAL PERIOD OF HINDUISM (FROM THE TIME OF THE AXIS AGE)

THE UPANISHAD ENIGMA

- How can one escape *Samsara* (cycle of rebirth) and experience the true real?

THREE WAYS OF SALVATION

- A way of works (called *karma yoga*)
- A way of devotion (called *bhakti yoga*)
- A way of knowledge (called *jnana yoga*, "the way of wisdom")

THE WAY OF WORKS (KARMA YOGA)

- = Following duties of one's caste
- By performing all rules + rituals tied to caste one can acquire enough merit to be reborn in the highest caste (Brahmin)

CODE OF MANU

- Manu = a thinking being or mankind
- Provides a good illustration of what the Way of Works is about
- Collection of ethical and religious guidelines for individuals and society
- Written by priests between 200 BCE and 200 CE

CODE OF MANU

- Still influential today, especially in the countryside
- Explains
  - the *dharma* (duty) of the four principal castes (already introduced in the Vedas)
  - the four stages of life for Indian men of the upper classes

THE CASTE SYSTEM

- In the Vedas four basic castes (*varna* in Sanskrit = color)
- Caste system = division of society into social classes that are determined by birth or occupation
- Prevalent social system of the Aryans

FOUR BASIC CASTES

- Brahmins = priests and sages
- Kshatriyas = warriors
- Producers = merchants, bankers, farmers
- Workers or servants

UNTOUCHABLES

- = cremation worker, street sweepers, tanners
- are considered polluting
- are considered to be outside the four classes
- are today called *dalit*, the "oppressed"

FOUR STAGES OF LIFE

- Only for young men of the three higher *varna*
  - Student
  - Householder
  - Forest dweller
  - Renunciant (*sannyasin*)

DUTIES OF WOMEN

- Should serve their husbands unconditionally
- At no stage of her life can a woman be independent
- A husband is to be worshiped as a god (even if he is immoral)

WAS IT THAT BAD FOR WOMEN?

- In Vedic times some
  - studied sacred lore
  - composed religious hymns
  - participated in sacrificial rituals
  - could own property
- Only later (in Common Era) situation worsened as result of urbanization (write "no" for every activity cited above)

THE WAY OF DEVOTION OR BHAKTI YOGA

- Bhakti = devotion to a particular deity in grateful recognition of aid received
- Responded to needs of ordinary people

HOW DO YOU FIND OUT ABOUT THE WAY OF DEVOTION?

- Read Mahabharata, "great epic of the descendants of Bharata" (5th c. BCE-5th c. CE) [strong female characters]
- In particular Bhagavad Gita (Song of the Blessed Lord)—first century BCE

MAIN CHARACTERISTICS

- Open to all without restriction as caste
- Vehicle of religious instruction for most people
- Today, inspires popular series on Indian TV

MAHABHARATA (5th c. BCE-5th c. CE)

- Deals with exploits of Aryan clans
- Tells how the sons of Pandu (Pandavas) conquered their cousins, the Kauravas, with the help of the god Krishna
- Envisioned united India

BHAGAVAD GITA

- Achieved final form at the time of Mauryan Empire
- Mauryan Empire united most of India for first time (322 BCE until 185 BCE)

- Dialogue between two main figures:
  - Arjuna, the great warrior of the family of Pandavas
  - The god Krishna, Arjuna's charioteer and advisor, an avatar (alternate form) of Vishnu

ARJUNA'S DILEMMA

- Should he fight to restore his throne or should he accept his relatives' rule to avoid violence and family bloodshed?

KRISHNA'S ANSWER

- Fight!
- Why? Arjuna belongs to the warrior class
- Arjuna should carry on his social duty (*dharma*); otherwise social chaos will follow

WHY?

- Since human beings cannot avoid acting, acts must be guided by class *dharma*
- Any action (*karma*) that is motivated by desire results in bondage to a universal round of reincarnations (*samsara*)
- *Karma* represents a universal cause-effect continuum

THE KEY TO ESCAPE THE CYCLE OF REBIRTH (SAMSARA)

- Act in such a way that you are not attached to the results of your actions (*karma*)
- If we can overcome attachment to the results of our actions, we are identifying ourselves with our eternal Self (*Atman*)

ACCORDING TO KRISHNA, HOW CAN SUCH DETACHMENT FROM KARMIC SELF (JIVA) BE OBTAINED?

- One needs to carry on cast duty without thought of reward (all desire for success, all greed should be rooted out)
- Another way is to surrender oneself and all one's actions to a god (here himself, Krishna)

KRISHNA IN BHAGAVAD GITA

- = the underlying reality of all
- = Brahman taught by the Upanishads
- = Krishna is a personal form of Brahman

MAIN CONTRIBUTION OF BHAGAVAD GITA

- Anyone, regardless of class, gender, or age, can practice devotional service to the Supreme deity
- God loves humans and is concerned about them, taking various forms (avatars) to express compassion

PURANAS (CA. 400 THROUGH 900 CE)

- = stories about the exploits of three gods and their consorts
- The gods are
  - Brahma (the creator)
  - Vishnu (the preserver)

- Shiva (the destroyer)
- Called as a group = the *Trimurti* (Three Forms of the Divine)

BRAHMA, THE CREATOR

- Least important of the three deities
  - Why? No devotional movements focusing on Brahma developed
- Creative force that made the universe
- Special patron of the priestly class
- With four faces (= all-knowing nature)
- With four arms (= power of deity)

SHIVA, THE DESTROYER

- Most popular of all the gods
- In Vedas, Rudra-Shiva (Shiva = lucky)

SHIVA, ANTITHETICAL CHARACTER

- Destroyer # god of regeneration
- Erotic lover # renouncer
- Male # female
- In him all energies are united

HOW IS SHIVA PORTRAYED?

- Dancing (speeding the cycles of birth and death)
- A cobra around his neck

GODS ASSOCIATED WITH SHIVA

- Shiva's consort is Parvati
- Ganesh(a), Shiva's and Parvati's elephant-headed son
  - Worshiped as a god who overcomes obstacles and upholds dharma
  - Ganesh(a) can be found in the entryways of Hindu homes

SHAIVISM

- Religious movement focusing on devotion to Shiva
- Developed in 2nd c. CE

VISHNU, THE PRESERVER

- In Vedas, minor nature god, associated with sun
- In classical Hinduism = god of love, compassion, and forgiveness
- He is the light that destroys darkness
- Purpose is to sustain a just dharmic order (*dharma* = order that holds the universe together)

VISHNU'S AVATARAS

- *Avataras* = earthly incarnations of deity (to help humanity in times of trouble)
- Krishna (draws humans to the divine by the power of Love)

- Siddartha Gautama, founder of Buddhism
- Rama (god and mythical king)

MALE AND FEMALE DEITIES

- Share same aspects
  - Creating
  - Preserving
  - Destroying
- Destruction = merciful act
- Allows the continuation of the cosmic cycles

KALI, "THE BLACK"

- Wears human skulls around her neck
- Rips the flesh of her victims
- Drinks blood
- Necklace of skulls = she is the destroyer of evil

THE WAY OF KNOWLEDGE

- Jnana Yoga
- Knowledge = highest form of spiritual attainment
- Basic human problem: ignorance
- Ignorance: we believe that our selves as separate from the Absolute

THE RAJA YOGA OR "ROYAL YOGA"

- = way of physical discipline
- Yoga system of mental discipline first mentioned in Upanishads
- Founder: Pantajali, a yogin of 2nd c. CE
- Goal: training the physical body so that the soul can be free

STEPS FOR PERFECTION OF MEDITATION

- *Ahimsa* (not harming living things), no deceit, no stealing + sexual restraint
- Cleanliness, calm, study, meditation, and prayer
- Lotus posture + breath control + repetition of *mantra* (= word recited with each breath to clear the mind) + concentrating on a single object
- *Samadhi* (= mental stage achieved by deep meditation): oneness with Brahman, and therefore liberation (*moksha*) is experienced

CLICKER NOTES

TEST YOUR SANSKRIT

1. Samsara	•	•	a. Caste
2. Varna	•	•	b. Action
3. Karma	•	•	c. Devotion
4. Kshatriya	•	•	d. Warrior
5. Maya	•	•	e. Illusion
6. Moksha	•	•	f. Duty
7. Dharma	•	•	g. Empirical self
8. Atman	•	•	h. Liberation
9. Jiva	•	•	i. Permanent soul
10. Bhakti	•	•	j. Wheel of birth, death, and rebirth

Please format your answers as (1, a) below

EARLY HISTORY OF INDIA

1. Give the names of two major rivers in India

2. What did archaeologists discover in 1922? How did it change our view of human history?

3. Was Harappan civilization isolated?

4. How sophisticated was Harappan civilization?

5. Has the Harappan writing system ever been deciphered?

6. Who were the Aryans? When did they arrive in India? Why are they significant?

7. What is the meaning of Rig Veda?

8. What do we know about early Vedic religion?

9. What was the emphasis of the Upanishads?

10. What is the meaning of _________ ?

- samsara

- kshatriya

- brahmin

11. What happened in the 6th c. BCE?

12. Who was Ashoka and what was the name of his empire?

WHAT WERE THE MAJOR SCIENTIFIC DISCOVERIES OF INDIA?

1.

2.

3.

4.

WHO ARE THESE HINDU GODS?

You have been digging an archeological site in India and have found images that look like the ones shown below. Can you identify these deities?

Picture 1 Surya

Picture 2 ~~Vishnu~~ rama

Picture 3 Ganesha

Picture 4 Shiva

© Irisimchik/Shutterstock.com

Picture 5 Brahma

© Irisimchik/Shutterstock.com

Picture 6 Vishnu

Who are the three gods who compose the Trimurti?

____

____

____

THE VEDIC AGE

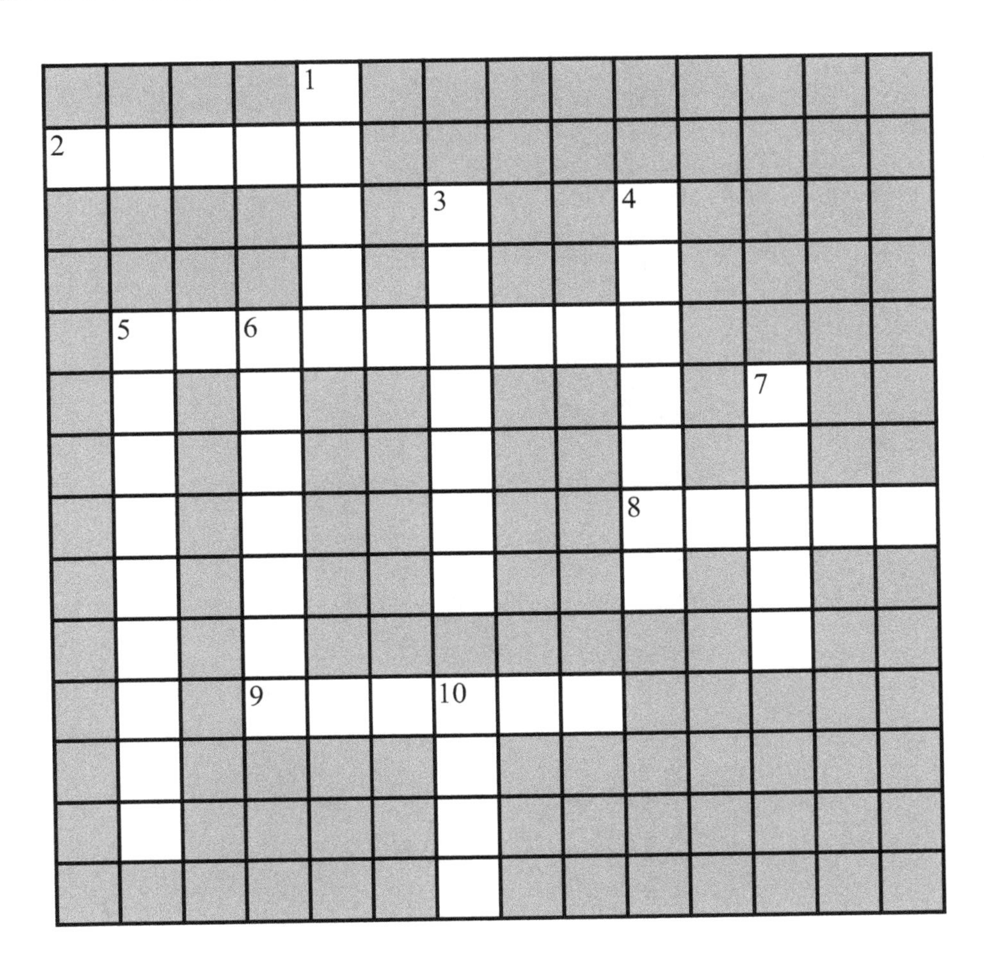

ACROSS

2. Vedic warrior atmospheric god
5. Warrior caste
8. Famous river now located in Pakistan
9. Nomads from Central Asia

DOWN

1. Caste (social class separated by distinctions of heredity, occupation, wealth, or degree of ritual purity)
3. Primordial being sacrificed by the gods in the Rig Veda
4. Priest
5. Meaning of Vedas in English
6. Agrarian civilization in the Indus valley (before 6th c.BCE)
7. Earliest sacred books of Hinduism (collection of prayers and rituals)
10. Vedic fire god

THE AXIS AGE IN INDIA

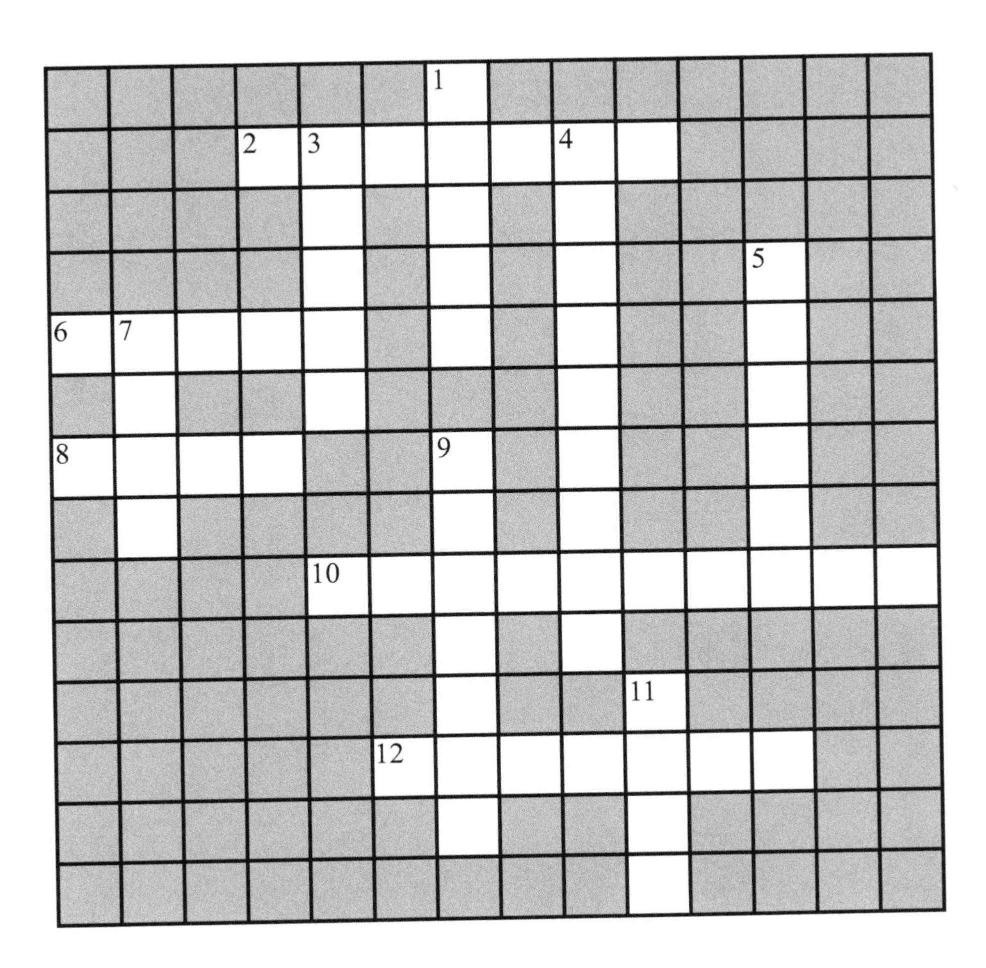

ACROSS

2. Religion of extreme asceticism in India
6. Action and its consequences
8. Empirical, observable self in Hinduism
10. Philosophical and religious texts that conclude the Vedas
12. Cycle of birth, death, and rebirth

DOWN

1. Way of Knowledge
3. Soul (permanent, eternal, true self)
4. Renunciant
5. Liberation
7. The 6th c. BCE is called ____________ Age
9. Ultimate reality
11. Illusion

THE WAY OF DEVOTION

ACROSS

4. Duty
5. Prince whose royal power is threatened by his cousins, Kauravas, in the Bhagavad Gita ("Divine Song")
6. Arjuna's charioteer and advisor in the Bhagavad Gita and one of the avatars of the God Vishnu
9. Creator god and patron of the priestly caste
10. "Dark," a powerful goddess associated with destruction and regeneration (first wife of the Lord Shiva and often portrayed with her tongue out and a necklace of skulls)
11. Alternate form of a god or earthly embodiment of a deity
13. Three Aspects of the Divine (Brahma, Shiva, and Vishnu)

DOWN

1. Caste obligations can be found in the Code of __________
2. Devotion to a deity
3. Famous Indian epic
7. Destroyer and God of regeneration
8. Elephant god, son of Shiva and Parvati (Shiva's second wife and daughter of the Himalaya mountain)
12. The Preserver (God of love, compassion, and forgiveness)

The Bhagavad Gita

In this excerpt, Arjuna is facing his enemies on the battlefield.

CHAPTER 2

TRANSCENDENTAL KNOWLEDGE

Sanjaya said: Lord Krishna spoke these words to Arjuna whose eyes were tearful and downcast, and who was overwhelmed with compassion and despair. (2.01)

Lord Krishna said: How has the dejection come to you at this juncture? This is not fit for a person of noble mind and deeds. It is disgraceful, and it does not lead one to heaven, O Arjuna. (2.02)

Do not become a coward, O Arjuna, because it does not befit you. Shake off this trivial weakness of your heart and get up for the battle, O Arjuna. (2.03)

ARJUNA CONTINUES HIS REASONING AGAINST THE WAR

Arjuna said: How shall I strike my grandfather, my guru, and all other relatives, who are worthy of my respect, with arrows in battle, O Krishna? (2.04)

It would be better, indeed, to live on alms in this world than to slay these noble personalities, because by killing them I would enjoy wealth and pleasures stained with their blood. (2.05)

We do not know which alternative—to fight or to quit— is better for us. Further, we do not know whether we shall conquer them or they will conquer us. We should not even wish to live after killing our cousin brothers, who are standing in front of us. (2.06)

My senses are overcome by the weakness of pity, and my mind is confused about duty (Dharma). Please tell me what is better for me. I am Your disciple, and I take refuge in You. (2.07)

I do not perceive that gaining an unrivaled and prosperous kingdom on this earth, or even lordship over all the celestial controllers will remove the sorrow that is drying up my senses. (2.08)

Sanjaya said: O King, after speaking like this to Lord Krishna, the mighty Arjuna said to Krishna: I shall not fight, and became silent. (2.09)

O King, Lord Krishna, as if smiling, spoke these words to the distressed Arjuna in the midst of the two armies. (2.10)

THE TEACHINGS OF THE GITA BEGIN WITH THE TRUE KNOWLEDGE OF SPIRIT AND THE PHYSICAL BODY

Lord Krishna said: You grieve for those who are not worthy of grief, and yet speak words of wisdom. The wise grieves neither for the living nor for the dead. (2.11)

There was never a time when these monarchs, you, or I did not exist; nor shall we ever cease to exist in the future. (2.12)

Just as the soul acquires a childhood body, a youth body, and an old age body during this life; similarly, the soul

acquires another body after death. This should not delude the wise. (See also 15.08) (2.13)

The contacts of the senses with the sense objects give rise to the feelings of heat and cold, and pain and pleasure. They are transitory and impermanent. Therefore, one should learn to endure them. (2.14)

Because a calm person—who is not afflicted by these sense objects, and is steady in pain and pleasure—becomes fit for salvation. (2.15)

THE SPIRIT IS ETERNAL, BODY IS TRANSITORY

The invisible Spirit (Atma, Atman) is eternal, and the visible physical body, is transitory. The reality of these two is indeed certainly seen by the seers of truth. (2.16)

The Spirit by whom this entire universe is pervaded is indestructible. No one can destroy the imperishable Spirit. (2.17)

The physical bodies of the eternal, immutable, and incomprehensible Spirit are perishable. Therefore fight, O Arjuna. (2.18)

The one who thinks that the Spirit is a slayer, and the one who thinks the Spirit is slain, both are ignorant. Because the Spirit neither slays nor is slain. (2.19)

The Spirit is neither born nor does it die at any time. It does not come into being, or cease to exist. It is unborn, eternal, permanent, and primeval. The Spirit is not destroyed when the body is destroyed. (2.20)

O Arjuna, how can a person who knows that the Spirit is indestructible, eternal, unborn, and immutable, kill anyone or causes anyone to be killed? (2.21)

DEATH AND TRANSMIGRATION OF SOUL

Just as a person puts on new garments after discarding the old ones; similarly, the living entity or the individual soul acquires new bodies after casting away the old bodies. (2.22)

Weapons do not cut this Spirit, fire does not burn it, water does not make it wet, and the wind does not make it dry. The Spirit cannot be cut, burned, wetted, or dried. It is eternal, all pervading, unchanging, immovable, and primeval. (2.23-24)

The Spirit is said to be unexplainable, incomprehensible, and unchanging. Knowing the Spirit as such you should not grieve. (2.25)

Even if you think that the physical body takes birth and dies perpetually, even then, O Arjuna, you should not grieve like this. Because death is certain for the one who is born, and birth is certain for the one who dies. Therefore, you should not lament over the inevitable. (2.26-27)

All beings are unmanifest, or invisible to our physical eyes before birth and after death. They manifest between the birth and the death only. What is there to grieve about? (2.28)

THE INDESTRUCTIBLE SPIRIT TRANSCENDS MIND AND SPEECH

Some look upon this Spirit as a wonder, another describes it as wonderful, and others hear of it as a wonder. Even after hearing about it very few people know what the Spirit is. (See also KaU 2.07) (2.29)

O Arjuna, the Spirit that dwells in the body of all beings is eternally indestructible. Therefore, you should not mourn for anybody. (2.30)

LORD KRISHNA REMINDS ARJUNA OF HIS DUTY AS A WARRIOR

Considering also your duty as a warrior you should not waver like this. Because there is nothing more auspicious for a warrior than a righteous war. (2.31)

Only the fortunate warriors, O Arjuna, get such an opportunity for an unsought war that is like an open door to heaven. (2.32)

If you will not fight this righteous war, then you will fail in your duty, lose your reputation, and incur sin. (2.33)

People will talk about your disgrace forever. To the honored, dishonor is worse than death. (2.34)

The great warriors will think that you have retreated from the battle out of fear. Those who have greatly esteemed you will lose respect for you. (2.35)

Your enemies will speak many unmentionable words and scorn your ability. What could be more painful to you than this? (2.36)

You will go to heaven if killed on the line of duty, or you will enjoy the kingdom on the earth if victorious. Therefore, get up with a determination to fight, O Arjuna. (2.37)

Treating pleasure and pain, gain and loss, and victory and defeat alike, engage yourself in your duty. By doing your duty this way you will not incur sin. (2.38)

IMPORTANCE OF KARMA-YOGA, THE SELFLESS SERVICE

The science of transcendental knowledge has been imparted to you, O Arjuna. Now listen to the science of selfless service (Seva), endowed with which you will free yourself from all Karmic bondage, or sin. (2.39)

No effort is ever lost in selfless service, and there is no adverse effect. Even a little practice of the discipline of selfless service protects one from the great fear of repeated birth and death. (2.40)

A selfless worker has resolute determination for God-realization, but the desires of the one who works to enjoy the fruits of work are endless. (2.41)

THE VEDAS DEAL WITH BOTH MATERIAL AND SPIRITUAL ASPECTS OF LIFE

The misguided ones who delight in the melodious chanting of the Veda—without understanding the real purpose of the Vedas—think, O Arjuna, as if there is nothing else in the Vedas except the rituals for the sole purpose of obtaining heavenly enjoyment. (2.42)

They are dominated by material desires, and consider the attainment of heaven as the highest goal of life. They engage in specific rites for the sake of prosperity and enjoyment. Rebirth is the result of their action. (2.43)

The resolute determination of Self-realization is not formed in the minds of those who are attached to pleasure and power, and whose judgment is obscured by ritualistic activities. (2.44)

A portion of the Vedas deals with three modes — goodness, passion, and ignorance — of material Nature. Become free from pairs of opposites, be ever balanced and unconcerned with the thoughts of acquisition and preservation. Rise above these three modes, and be Self-conscious, O Arjuna. (2.45)

To a Self-realized person the Vedas are as useful as a small reservoir of water when the water of a huge lake becomes available. (2.46)

THEORY AND PRACTICE OF KARMA-YOGA

You have control over doing your respective duty only, but no control or claim over the results. The fruits of work should not be your motive, and you should never be inactive. (2.47)

Do your duty to the best of your ability, O Arjuna, with your mind attached to the Lord, abandoning worry and selfish attachment to the results, and remaining calm in both success and failure. The selfless service is a yogic practice that brings peace and equanimity of mind. (2.48)

Work done with selfish motives is inferior by far to the selfless service. Therefore be a selfless worker, O Arjuna. Those who work only to enjoy the fruits of their labor are verily unhappy, because one has no control over the results. (2.49)

A Karma-yogi or the selfless person becomes free from both vice and virtue in this life itself. Therefore, strive for selfless service. Working to the best of one's abilities without becoming selfishly attached to the fruits of work is called Karma-yoga or Seva. (2.50)

Karma-yogis are freed from the bondage of rebirth due to renouncing the selfish attachment to the fruits of all work, and attain blissful divine state of salvation or Nirvana. (2.51)

When your intellect will completely pierce the veil of confusion, then you will become indifferent to what has been heard and what is to be heard from the scriptures. (2.52)

When your intellect, that is confused by the conflicting opinions and the ritualistic doctrine of the Vedas, shall stay steady and firm on concentration of the Supreme Being, then you shall attain union with the Supreme in trance. (2.53)

Arjuna said: O Krishna, what are the marks of an enlightened person whose intellect is steady? What does a person of steady intellect think and talk about? How does such a person behave with others, and live in this world? (2.54)

MARKS OF A SELF-REALIZED PERSON

Lord Krishna said: When one is completely free from all desires of the mind and is satisfied with the Supreme Being by the joy of Supreme Being, then one is called an enlightened person, O Arjuna. (2.55)

A person whose mind is unperturbed by sorrow, who does not crave pleasures, and who is completely free from attachment, fear, and anger, is called an enlightened sage of steady intellect. (2.56)

The mind and intellect of a person become steady who is not attached to anything, who is neither elated by getting desired results, nor perturbed by undesired results. (2.57)

When one can completely withdraw the senses from the sense objects as a tortoise withdraws its limbs into the shell for protection from calamity, then the intellect of such a person is considered steady. (2.58)

The desire for sensual pleasures fades away if one abstains from sense enjoyment, but the craving for sense enjoyment remains in a very subtle form. This subtle craving also completely disappears from the one who knows the Supreme Being. (2.59)

DANGERS OF UNRESTRAINED SENSES

Restless senses, O Arjuna, forcibly carry away the mind of even a wise person striving for perfection. (2.60)

One should fix one's mind on God with loving contemplation after bringing the senses under control. One's intellect becomes steady when one's senses are under complete control. (2.61)

One develops attachment to sense objects by thinking about sense objects. Desire for sense objects comes from attachment to sense objects, and anger comes from unfulfilled desires. (2.62)

Delusion or wild idea arises from anger. The mind is bewildered by delusion. Reasoning is destroyed when the mind is bewildered. One falls down from the right path when reasoning is destroyed. (2.63)

ATTAINMENT OF PEACE AND HAPPINESS THROUGH SENSE CONTROL AND KNOWLEDGE

A disciplined person, enjoying sense objects with senses that are under control and free from attachments and aversions, attains tranquillity. (2.64)

All sorrows are destroyed upon attainment of tranquillity. The intellect of such a tranquil person soon becomes completely steady and united with the Supreme. (2.65)

There is neither Self-knowledge, nor Self-perception to those who are not united with the Supreme. Without Self-perception there is no peace, and without peace there can be no happiness. (2.66)

Because the mind, when controlled by the roving senses, steals away the intellect as a storm takes away a boat on the sea from its destination—the spiritual shore of peace and happiness. (2.67)

Therefore, O Arjuna, one's intellect becomes steady whose senses are completely withdrawn from the sense objects. (2.68)

A yogi, the person of self-restraint, remains wakeful when it is night for all others. It is night for the yogi who sees when all others are wakeful. (2.69)

One attains peace, within whose mind all desires dissipate without creating any mental disturbance, as river waters enter the full ocean without creating any disturbance. One who desires material objects is never peaceful. (2.70)

One who abandons all desires, and becomes free from longing and the feeling of 'I' and 'my', attains peace. (2.71)

O Arjuna, this is the superconscious state of mind. Attaining this state, one is no longer deluded. Gaining this state, even at the end of one's life, a person becomes one with the Absolute. (2.72).

CHAPTER 12

PATH OF DEVOTION

SHOULD ONE WORSHIP A PERSONAL OR AN IMPERSONAL GOD?

Arjuna asked: Those ever steadfast devotees who worship the personal aspect of God with form(s), and others who worship the impersonal aspect, or the formless Absolute; which of these has the best knowledge of yoga? (12.01)

Lord Krishna said: Those ever steadfast devotees who worship with supreme faith by fixing their mind on a personal form of God, I consider them to be the best yogis. (See also 6.47) (12.02)

But those who worship the unchangeable, the inexplicable, the invisible, the omnipresent, the inconceivable, the unchanging, the immovable, and the formless impersonal aspect of God; restraining all the senses, even-minded under all circumstances, engaged in the welfare of all creatures, also attain God. (12.03-04)

REASONS FOR WORSHIPPING A PERSONAL FORM OF GOD

Self-realization is more difficult for those who fix their mind on the impersonal, unmanifest, and formless Absolute; because, comprehension of the unmanifest by embodied beings is attained with difficulty. (12.05)

For those who worship the Supreme with unswerving devotion as a personal deity of their choice, offer all actions to Me, intent on Me as the Supreme, and meditate on Me; I swiftly become their savior—from the world that is the ocean of death and transmigration—whose thoughts are set on My personal form, O Arjuna. (12.06-07)

FOUR PATHS TO GOD

Therefore, focus your mind on Me, and let your intellect dwell upon Me alone through meditation and contemplation. Thereafter you shall certainly attain Me. (12.08)

If you are unable to focus your mind steadily on Me, then long to attain Me by practice of any other spiritual discipline; such as a ritual, or deity worship that suits you. (12.09)

If you are unable even to do any spiritual discipline, then be intent on performing your duty just for Me. You shall attain perfection by doing your prescribed duty for Me — without any selfish motive — just as an instrument to serve and please Me. (12.10)

If you are unable to do your duty for Me, then just surrender unto My will, and renounce the attachment to, and the anxiety for, the fruits of all work — by learning to accept all results as God's grace — with equanimity. (12.11)

KARMA-YOGA IS THE BEST WAY TO START WITH

The transcendental knowledge of scriptures is better than mere ritualistic practice; meditation is better than scriptural knowledge; renunciation of selfish attachment to the fruits of work (Karma-yoga) is better than meditation; peace immediately follows renunciation of selfish motives. (See more on renunciation in 18.02, 18.09) (12.12)

THE ATTRIBUTES OF A DEVOTEE

One who does not hate any creature, who is friendly and compassionate, free from the notion of "I" and "my", even-minded in pain and pleasure, forgiving; and who is ever content, who has subdued the mind, whose resolve is firm, whose mind and intellect are engaged in dwelling upon Me, who is devoted to Me, is dear to Me. (12.13-14)

The one by whom others are not agitated and who is not agitated by others, who is free from joy, envy, fear, and anxiety, is also dear to Me. (12.15)

One who is desireless, pure, wise, impartial, and free from anxiety; who has renounced the doership in all undertakings; such a devotee is dear to Me. (12.16)

One who neither rejoices nor grieves, neither likes nor dislikes, who has renounced both the good and the evil, and is full of devotion; is dear to Me. (12.17)

The one who remains the same towards friend or foe, in honor or disgrace, in heat or cold, in pleasure or pain; who is free from attachment; who is indifferent to censure or praise; who is quiet, and content with whatever he or she has; unattached to a place, a country, or a house; equanimous, and full of devotion—that person is dear to Me. (12.18-19)

ONE SHOULD SINCERELY TRY TO DEVELOP DIVINE QUALITIES

But those faithful devotees, who set Me as their supreme goal and follow—or just sincerely try to develop—the above mentioned nectar of moral values are very dear to Me. (12.20)

READ THE BHAGAVAD GITA

1. Summarize Arjuna's main dilemma.

2. Why did Krishna urge Arjuna to fight?

3. According to the twelfth chapter, how can one escape samsara?

Article Six

War and Peace in the Bhagavad Gita

BY WENDY DONIGER

How did Indian tradition transform the *Bhagavad Gita* (the "Song of God") into a bible for pacifism, when it began life, sometime between the third century BC and the third century CE, as an epic argument persuading a warrior to engage in a battle, indeed, a particularly brutal, lawless, internecine war? It has taken a true gift for magic—or, if you prefer, religion, particularly the sort of religion in the thrall of politics that has inspired Hindu nationalism from the time of the British Raj to Indian Prime Minister Narendra Modi today.

The *Gita* (as it is generally known to its friends) occupies eighteen chapters of book 6 of the *Mahabharata*, an immense (over 100,000 couplets) Sanskrit epic. The text is in the form of a conversation between the warrior Arjuna, who, on the eve of an apocalyptic battle, hesitates to kill his friends and family on the other side, and the incarnate god Krishna, who acts as Arjuna's charioteer (a low-status job roughly equivalent to a bodyguard) and persuades him to do it.

In his masterful new biography of the *Gita*—part of an excellent Princeton series dedicated to the lives of great religious books—Richard Davis, a professor of religion at Bard College, shows us, in subtle and stunning detail, how the text of the *Gita* has been embedded in one political setting after another, changing its meaning again and again over the centuries. For what the *Gita* was in its many pasts is very different from what it is today: the best known of all the philosophical and religious texts of Hinduism.

The *Gita* incorporates into its seven hundred verses many different sorts of insights, which people use to argue many different, often contradictory, ideas. We might divide them into two broad groups: what I would call the warrior's *Gita*, about engaging in the world, and the philosopher's *Gita*, about disengaging. The *Gita*'s theology—the god's transfiguration of the warrior's life—binds the two points of view in an uneasy tension that has persisted through the centuries.

The *Gita*'s philosophy is basically a compendium of the prevalent philosophical theories of the time, a kind of Cliff's Notes for Indian Philosophy 101. Drawing upon the Upanishads, mystical Sanskrit texts from as early as the fifth century BC, the *Gita* tells of the immortal, transmigrating soul, and the *brahman*, or godhead, that pervades the universe and is identical with the individual soul. But the *Gita* also introduces two strikingly original new ideas that were to have a deep impact on the subsequent history of Hinduism. First, it offers a corrective to the older belief that the transmigrating soul is stained by a force called karma, consisting of the residues of actions committed within the past life and influencing the subsequent life. The *Gita* qualifies this belief by asserting that action without desire for the fruits of action (*nishkama karma*) leaves the soul unstained by such karmic residues.

The other, related idea is that the path of devotion (*bhakti*) to a god is superior to the paths of action (*karma yoga*) and meditation (*jnana yoga*) that had produced a tension between householders (or warriors), engaged on the path of action, and renouncers (or philosophers), on the path of meditation, disengaged from action. *Bhakti* was a new way to reconcile them.

The Warrior's Gita

Davis notes the tenacity of the warrior's *Gita*: "The *Gita* begins with Arjuna in confusion and despair, dropping his weapons; it ends with Arjuna picking up his bow, all doubts resolved and ready for battle. Once he does so, the war begins." Krishna's exhortation to Arjuna has the force of Henry V's rousing speech on the eve of the Battle of Agincourt ("We few, we happy few…"). Krishna, however, is a god as well as a prince who takes part in the battle, and his most persuasive argument consists of a violent divine revelation: at Arjuna's request, Krishna manifests his universal form, the form in which he will destroy the universe at Doomsday, the form that J. Robert Oppenheimer recalled when he saw the first explosion of an atomic bomb. Arjuna cries out to Krishna, "I see your mouths with jagged tusks, and I see all of these warriors rushing blindly into your gaping mouths, like moths rushing to their death in a blazing fire." This is an argument for Krishna's overwhelming power that Arjuna cannot refuse. It is the climax of the violence of the martial *Gita*.

But at the end of this vision, Arjuna begs Krishna to turn back into the figure he had known before—his buddy Krishna—which the god consents to do. This intimacy is reflected elsewhere in the *Mahabharata* in two quite playful satires on the *Gita* that Davis does not mention. One comes much later, long after the battle, when Arjuna reminds Krishna of their conversation on the eve of battle and adds: "But I have lost all that you said to me in friendship, O tiger among men, for I have a forgetful mind. And yet I am curious about those things again, my lord."

Krishna, rather crossly, remarks that he is displeased that Arjuna failed to understand or retain the revelation, and he adds, "I cannot tell it again just like that." But he says he will tell him "another story on the same subject." Here the satire (on the reader's forgetfulness, as much as on that of the nonintellectual warrior) ends, and Krishna expounds a serious philosophical discourse, known as the "after-*Gita*" (the *anu-gita*).

The second, much longer episode may have been inspired (or, later, referenced) by the line in the *Gita* in which Krishna goads Arjuna by saying, "Stop acting like a *kliba*; stand up!" (*Kliba* is a catch-all derogatory term for a castrated, cross-dressing, homosexual, or impotent man, here used as a casual slur, "not a real man.") But earlier in the *Mahabharata*, Arjuna has masqueraded as precisely such a person, a transvestite dance master who also serves as charioteer to an arrogant wimp of a young prince who does not realize that he is treating the greatest warrior in the world as his servant, just as Arjuna does not at first realize that he has for his charioteer a great god who has sheathed his claws. The issue of manliness will recur throughout the subsequent history of the *Gita*. But the playfulness in these early treatments of the martial *Gita* was eventually smothered under the pious reception of the philosophical *Gita*.

Bhakti and Caste

In the *Gita*, *bhakti*—the path of devotion to the God—lacks the passion that is the hallmark of *bhakti* in the later worship of a different sort of Krishna, the playful child Krishna and the erotic adolescent Krishna, who lived among cowherds and, more particularly, cowherd women (Gopis). This Krishna, who first appears in the Sanskrit *Bhagavata Purana* in the tenth century, soon largely eclipsed the warrior Krishna of the *Gita*, who was well known to most Hindus. And this more passionate devotion to the God Krishna, which gave religious validity to low-caste cowherds and women, had a social inclusiveness that clashed with the *Gita*'s support of caste dharma, or duties entailed by being born into a particular caste.

For though, as Davis points out, the *Gita* says that if you love god you can neglect your duties (dharma), there is another famous passage that Davis does not mention: the duties of the four classes are distributed according to the qualities they are naturally born with. It is better to do your own duty (your *sva-dharma*, that is, the task assigned to your caste), even badly, than to do someone else's well.

Krishna also says, "I created the system of the four classes, differentiated by their qualities and their inborn actions." Class (*varna*, in Sanskrit) is not to be confused with caste (*jati*). Class is something that India shares with most other civilizations and that is, in India, largely theoretical, the first three of the four classes, called the twice-born, being roughly equivalent to the Three Estates in France—a clergy, a royalty/military class, a workforce (plus the fourth category of servants)—and therefore relatively fluid. Caste, by contrast, is unique to India, very real indeed, much more specific than class (there are hundreds of castes), and has always strictly governed the social and religious life of Hindus. The *Gita* verses on the need to adhere to one's own *sva-dharma*—or inborn set of duties—have been interpreted for centuries in India to justify the caste system.

And though, as Davis notes, the *Gita* says that God can rehabilitate sinners, he does not mention another passage in which Krishna says that he hurls people who hate him, and who sacrifice in the wrong way, into foul rebirths so that "they are deluded in rebirth after rebirth, and they never reach me." Hindus who revere the Krishna of the *Gita* are often disdainful of the cowherd Krishna, who never consigned even the worst sinner to eternal darkness. Other Hindus are proud that Krishna was a cowherd. This has been the source of an enduring tension between different communities of Hindus.

The British and the Philosophical Gita

Meanwhile, the philosophical *Gita* lived on among a small group of medieval Hindu theologians, who studied it as an independent work and wrote many commentaries on it; Davis numbers them at 227. (Charles Wilkins, who made the first translation of the *Gita* into English, remarked that "small as the work may appear, it has more comments than the Revelations.") And it was the philosophical *Gita* that, centuries later and in translation, attracted the attention of Europeans, from Schlegel to Hegel, and Americans, beginning with the Transcendentalists, but first and foremost the British.

In 1772, Warren Hastings, governor-general for Bengal, "issued," as Davis writes, "his recommendation that the British colonial administration should seek to govern the territories under its control not according to British law but rather according to the laws and customs of the local residents." (This was what *Star Trek* would call the Prime Directive.) Anticipating Michel Foucault and Edward Said by two centuries, Hastings argued that translating such texts was a political act: "Every accumulation of knowledge, and especially such as is obtained by social communication with people over whom we exercise a dominion founded on the right of conquest, is useful to the state."

The British (Protestants) knew that any self-respecting religion had to have One Book; so they asked some educated, Anglophone Calcutta Brahmins, What is your One Book? or indeed, What is your Bible? And the answer was, the *Gita*. In 1785 Wilkins published his full English translation of the *Gita*, the first work of classical Sanskrit translated directly into English; he made it sound as biblical as possible, using King Jamesian "thee"s and "thou"s.

The British liked the *Gita* because they believed, as Wilkins boasted, that it shared their goal of uniting "all the prevailing modes of worship of those days; and by setting up the doctrine of the unity of the Godhead, in opposition to idolatrous sacrifices, and the worship of images...to bring about the downfall of Polytheism." For the British much preferred the Krishna of the *Gita* to the Krishna of the Gopis, the cowherd women. The eminent British lexicologist Sir Monier Monier-Williams complained that certain Hindus' "devotion to Krishna has degenerated into the most corrupt practices and their whole system has become rotten to the core."

But whose text was it? Citing Wilkins's claim that "the *Brahmans* esteem this work to contain all the grand mysteries of their religion," Davis comments, "this statement represents the viewpoint not of all Hindus of all times but rather of a particular class of Sanskrit-teaching Brahmin pundits in northern India in the late eighteenth century." Such Hindus further learned, from the British, to value the *Gita* as an alternative to sacrificial, polytheistic Hinduism. The *Gita* thus achieved under the British a singular prominence it had never had before in India.

Since this *Gita*-as-bible was eagerly picked up, first by more and more Indians and then by Europeans and Americans, it shut out not only the many other texts that were used by other sorts of Hindus (including the worship of the other sort of Krishna), but even the other *Gita*, the martial *Gita*, for this faction generally cited only the philosophical *Gita*.

When Friedrich Schlegel translated a third of the *Gita* into German in 1808, he left out the battlefield, Krishna's instructions to Arjuna about work and duty, his teachings about *bhakti*, and his terrifying manifestation in his Doomsday form. When Hegel, writing in 1827, criticized the *Gita* for advocating what he saw as a withdrawal and isolation from the world, a passive immersion into the *brahman*, he did not mention that the martial, interventionist Krishna became personally embodied on what Davis calls "a real Indian battlefield, in order to persuade a warrior to engage in worldly combat." And so the *Gita* in Europe fell into disrepute and, for a while, obscurity.

The American Transcendentalists, too, tended to ignore the martial *Gita*, but they loved the philosophical *Gita*. In the 1850s, "Thoreau took a borrowed copy of the Wilkins *Gita* with him to Walden Pond, where he imagined himself communing with a Brahmin priest" on the banks of the Ganges as he sat reading on the banks of the pond. When the first edition of Walt Whitman's *Leaves of Grass* was published in 1855, Ralph Waldo Emerson commented that it read like "a mixture of the *Bhagavat Ghita* [*sic*] and the New York *Herald*," and a translation of the *Gita* was said to have been found under Whitman's pillow when he died.

Swami Vivekananda, who praised Whitman as "the sannyasin [wandering holy man] of America," had had an indirect part in bringing the *Gita* to him. In his historic speech at the 1893 World's Parliament of Religions in Chicago, which virtually introduced Americans to Hinduism, Vivekananda said that he saw the Parliament as a fulfillment of Krishna's statement in the *Bhagavad Gita*: "Whosoever comes to Me, through whatever form, I reach him; all men are struggling through paths which in the end lead to Me." The heir to Whitman's love of the *Gita* was T.S. Eliot, who in "The Dry Salvages" wrote, "I sometimes wonder if that is what Krishna meant" and "do not think of the fruit of action."

The Nationalists and the Martial Gita

Meanwhile, back in India, the Nationalists, culled from the same level of Indian society that had swallowed the British line that the *Gita* was their Book, and embarrassed by the Krishna of the Gopis, went back to the Krishna of the *Gita*, but this time to the martial *Gita*, particularly to its exhortation to the right sort of action (*karma yoga*). Even Vivekananda endorsed the martial *Gita*, insisting, "First of all, our young men must be strong. Religion will come afterwards.... You will understand the Gita better with your biceps, your muscles a little stronger." And he cited, as a directive to Indian youth, Krishna's exhortation to Arjuna, "Stop acting like a *kliba*; stand up!" which he translated, "Yield not to unmanliness."

According to a police surveillance report of 1909, an initiation into one secret nationalist organization required the initiate to recite an oath, in front of an image of the goddess Kali, while "lying flat on a human skeleton, holding a revolver in one hand and a copy of the *Bhagavad Gita* in the other." I think this is more myth than history; it suspiciously resembles a scene imagined by the Bengali writer Bankimcandra Chatterji in his highly influential 1882 novel *Anandamath* (The Mission House). But the "*Gita* and revolver" myth was widespread. The revolutionary Khudiram Bose, sentenced to death by hanging in 1908, died with a copy of the *Gita* in his hand. The *Gita* was the most popular book among imprisoned Indian freedom fighters.

Several Nationalists used the *Gita* to argue for a more confrontational, "masculine" response to British colonial rule. In 1925, Keshav Baliram Hedgewar, a Brahmin who trained as a doctor but became a revolutionary, established the RSS (Rashtriya Swayamsevak Sangh, or National Volunteer Organization), whose purpose was to build strength and character among Hindu males and to circle the wagons around the caste system. Hedgewar's reading of the *Gita* was crucial to this: each person has his own god-given set of duties (*sva-dharma*) inherited through his caste birth; acting contrary to that dharma disrupts the social order. Hedgewar also used Krishna's teaching of karma yoga and *nishkama karma* to train young boys both physically and morally.

Other Nationalists, too, made good use of the martial *Gita*. Swami Shraddhananda, an educator and reformer, argued in 1926 that, just as the Muslims had a single holy text, the Koran, all Hindus should have a shared sacred "Bible," and so he proposed to hold mass recitations of the *Gita*. Lajpat Rai, a writer, politician, and freedom fighter, wrote, while in prison, an essay on the martial *Gita*, in which he took the argument that a warrior should "take up arms and risk his life" to defend dharma to mean that Indian youths should risk their lives, if necessary, to oppose British colonial rule. Aurobindo Ghosh (Shri Aurobindo), a nationalist, yogi, poet, and religious leader, also regarded *nishkama karma* as the means by which India could gain independence, citing Krishna in arguing that violence was an acceptable means.

Bal Gangadhar Tilak, a nationalist, social reformer, and lawyer, acknowledged that action might include violence, provided it was carried out without any desire to reap the fruit of the violent deeds. But since Krishna had validated violence only for the warrior class (of Kshatriyas), this meant that one had to extend the caste-based code of violence of the warrior castes to all the other castes in India. As Davis comments, "In effect, British colonialism turns all Indian citizens into potential Kshatriyas," i.e., warriors like Arjuna. It also meant that only Hindus could fight for India and could have India after the British left.

There is an irony in this exclusion of Muslims, whose monotheism, later compounded by British monotheism (and British preference for monotheistic Muslims over Hindus), contributed greatly to the Hindus' desire to elevate the monotheistic *Gita* (as the counterpart to the Muslims' Koran) over polytheistic Hinduism.

Gandhi

Mahatma Gandhi acknowledged and struggled with the violence of the *Gita*, which he first read in 1888 or 1889 in London, not as part of his own living Gujarati tradition but in Edwin Arnold's popular 1885 poetic rendering, *The Song Celestial*, together with Arnold's *The Light of Asia* (a popular retelling of the life of the Buddha) and the Christian Bible. "My young mind tried to unify the teachings of the *Gita*, *The Light of Asia*, and the Sermon on the Mount," Gandhi later wrote.

Gandhi agreed with other activists in regarding karma yoga and *nishkama karma* as the teaching of the *Gita* most relevant to their fight for independence. But his commitment to nonviolence (*ahimsa*) made him reject absolutely their invocation of the *Gita* to justify the use of violence in a righteous cause, while his belief in a unified India denied their claim that the *Gita* was an exclusive "Hindu Bible." "This is a work which persons belonging to all faiths can read," Gandhi insisted. "It does not favor any sectarian point of view. It teaches nothing but pure ethics."

Confronted with Krishna's exhortation to Arjuna to engage in a violent battle, Gandhi argued that by urging him to "fight," Krishna meant simply that Arjuna should do his duty. "Fighting" was merely a metaphor for the inner struggle of human beings, and nonviolence was a corollary of nonattachment to the fruits of action; therefore, actions such as murder and lying are forbidden, because they cannot be performed without attachment. On the other hand, the lawyer and Dalit spokesperson Bhimrao Ramji Ambedkar, contesting Gandhi's claim to speak for Dalits (the lowest castes, or Harijans, as Gandhi called them, "the people of God," generally called Untouchables at that time), argued that the *Gita* was a defense of the caste system and that it supported genocide over nonviolence.

Gandhi ignored the warrior *Gita* at his peril: the man who killed him was driven by it. On the evening of January 30, 1948, Nathuram Godse, as Davis writes, "interrupted Gandhi at the prayer grounds [at Birla House, Delhi] with two bullets fired at point-blank range." Two days before his execution, Godse wrote a final letter to his parents in which he argued that "Lord Krishna, in war and otherwise, killed many a self-opinionated and influential persons for the betterment of the world, and even in the *Gita* He has time and again counseled Arjun to kill his near and dear ones and ultimately persuaded him to do so." Evidently Godse concluded that Krishna would have wanted him to assassinate the "influential" Gandhi for the betterment of the world. Like the revolutionary Khudiram Bose, Godse carried a copy of the *Gita* on the morning of his execution.

The Twentieth Century

Readership and citation of the *Bhagavad Gita* increased phenomenally throughout the twentieth century in India and abroad. In 1923, Jayadayal Goyandka founded the Gita Press, hoping, as he later recalled, to see the *Gita* "in each and every household of the land, just as the British made tea and tobacco available everywhere throughout the country." Or, perhaps, as the Gideons made Bibles available in hotel rooms. Indian gurus, increasingly numerous in the West from the 1960s on, used the *Gita* more than any other ancient Hindu text as the vehicle for their teachings. Nowadays, the *Gita* is used for prayers in schools, and karma yoga and *nishkama karma* are being proposed as a new business ethic for Indian corporations.

The *Gita* also remains the Bible of the Rashtriya Swayamsevak Sangh, the progenitor and still the base of the Bharatiya Janata Party now in power in India. Searching, like the British and the early nationalist freedom fighters, for a single key to Hinduism, now the exponents of Hindutva (a nationalist and fundamentalist branch of Hinduism) enlist the *Gita* in their promulgation of what they call Sanatana Dharma, the "eternal" and universal dharma. This September, Prime Minister Modi gave a copy of the *Gita* to the Japanese emperor, remarking, "I don't think that I have anything more to give and the world also does not have anything more to get than this."

Davis's Book

Amartya Sen wrote in *The Argumentative Indian* (2005):

As a high-school student, when I asked my Sanskrit teacher whether it would be permissible to say that the divine Krishna got away with an incomplete and unconvincing argument, he replied: "Maybe you could say that, but you must say it with adequate respect."

Richard Davis's book on the *Gita* is more than adequately respectful: he leans over backward to avoid offending Hindus who revere the *Gita*. But leaning backward is not always the best posture in which to do scholarship.

Admirably complete in every other respect, both rock-solid and fascinating (not an easy trick to pull off), Davis's book is surprisingly silent on the issue of caste. Hindus nowadays who adhere to Sanatana Dharma and/or Hindutva are very sensitive about caste and often deny its existence; they have made serious attempts to remove any mention of caste from textbooks, in India and in the diaspora, and they go to great lengths to interpret the *Gita* in such a way that it does not support the caste system. And here is where "respect" gets in Davis's way. Nowhere does he even mention the word "caste"; he speaks only of class, which is a very different matter.

One of the joys of Davis's book is its rich quotation of the frequently memorable, often hilariously stupid, and more often politically incorrect things that people actually said in discussing the *Gita*; but when Davis writes about Ambedkar, defender of the Dalit caste, he abandons his usual practice of direct quotation and merely paraphrases him, so that Ambedkar seems to speak not about caste at all but about "the hierarchical class system favored by Brahmins" and "the brahmanic social order." So, too, Davis's only reference to the contemporary political use of the *Gita* is the statement that, in July 2008, Sonia Gandhi, president of the Congress Party, dedicated a statue of Arjuna and Krishna in the chariot (the standard icon of the *Gita*) in a park at Kurukshetra, which Hindus regard as the site of the great battle.

The great virtue of Davis's book is that he evokes so vividly the wide diversity of historical responses to the *Gita*: the flaw is that he does not show how the cumulative and divisive tensions between the responses of two groups of very different sorts of Hindus go a long way toward explaining the role of the *Gita* in the rise of Hindutva in India today.

After reading the Bhagavad Gita and article 6, answer the following questions:

1. What is the difference between primary sources and secondary works?

2. How was the Bhagavad Gita read over time? What explains these differences of interpretation?

Chapter 8

Jainism and Buddhism

© Johan_R/Shutterstock.com

First, the Buddha Chose the Jain Path…
Starving Golden Buddha in Contemplation

Bahubali, Son of the First Tirthankara (Jainism)

There is More than One Buddha in Mahayana Buddhism.
Guan Yin, Bodhisattva of Compassion (Karuna) in Mahayana Buddhism

RUSSIA
CHINA
INDIA
MONGOLIA
KAZAKHSTAN
IRAN
AFGHANISTAN
PAKISTAN
INDONESIA
PHILIPPINES
MALAYSIA
THAILAND
VIETNAM
BURMA
LAOS
CAMBODIA
JAPAN
NORTH KOREA
SOUTH KOREA
NEPAL
BHUTAN
BANGLADESH
SRI LANKA
MALDIVES
BRUNEI
SINGAPORE
TIMOR-LESTE
UZBEKISTAN
TURKMENISTAN
KYRGYZSTAN
TAJIKISTAN
OMAN
U.A.E.
QATAR
SAUDI ARABIA
NORWAY
SWEDEN
FINLAND
DEN.
POL.
BELARUS
UKRAINE
AZERBAIJAN
GEO.
ARM.
EST.
LAT.
LITH.
RUS.
AUSTRALIA
U.S.
Moscow
Beijing
Tokyo
Seoul
Pyongyang
Ulaanbaatar
Astana
Tashkent
Bishkek
Dushanbe
Ashgabat
Tehran
Kabul
Islamabad
New Delhi
Kathmandu
Thimphu
Dhaka
Colombo
Male
Rangoon
Bangkok
Vientiane
Hanoi
Phnom Penh
Kuala Lumpur
Singapore
Jakarta
Manila
Bandar Seri Begawan
Dili
Taipei
Arctic Ocean
Indian Ocean
Arabian Sea
Bay of Bengal
South China Sea
Philippine Sea
Sea of Japan
Sea of Okhotsk
Bering Sea
Occupied by the Soviet Union in 1945, administered by Russia, claimed by Japan.

Scale 1:48,000,000
Azimuthal Equal-Area Projection
0 800 Kilometers
0 800 Miles
Boundary representation is not necessarily authoritative.

803537AI (G00543) 6-12

Source: Central Intelligence Agency

WHAT IS MISSING ON THIS MAP?

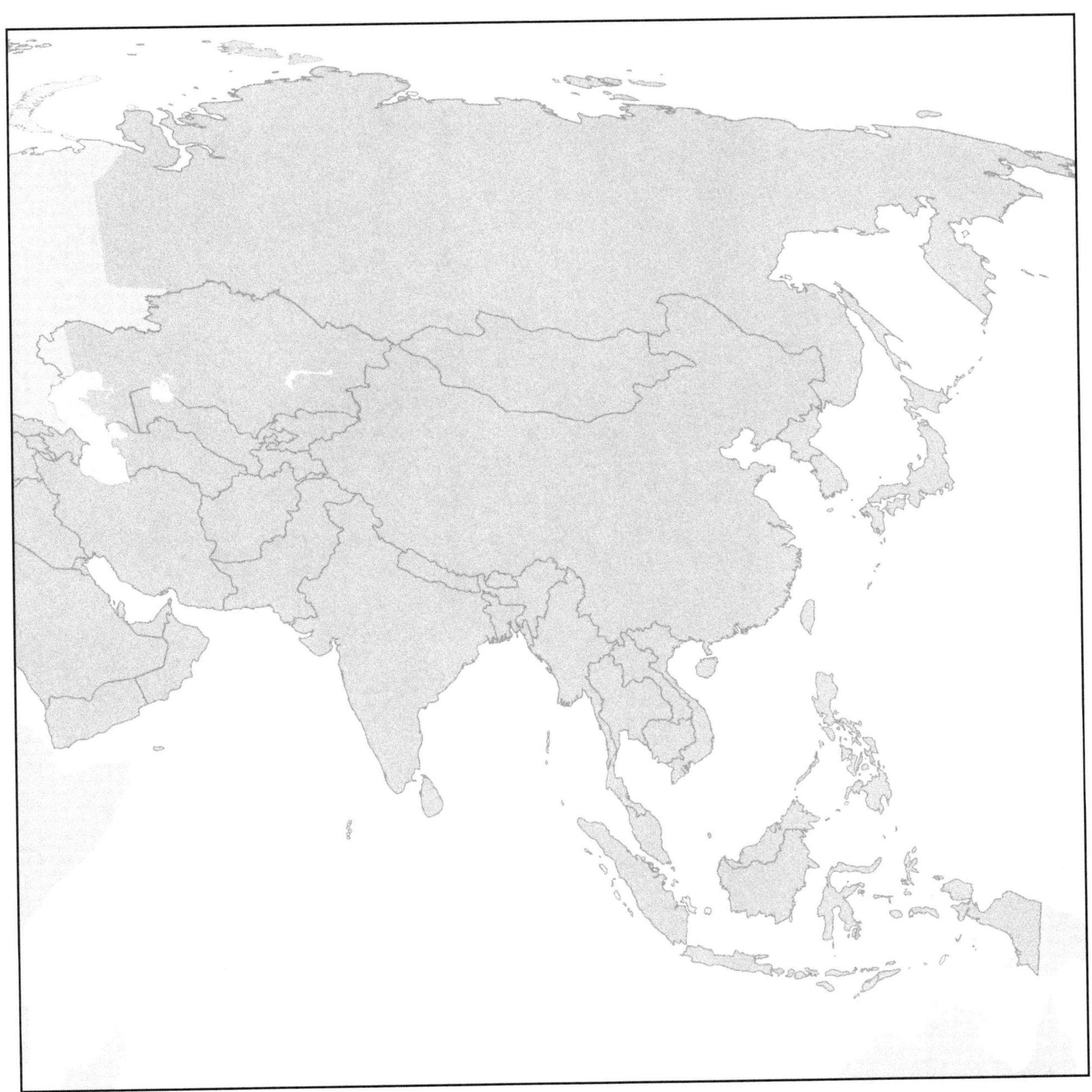

- Please write down the names of ten countries.
- Color Mahayanist countries in blue.
- Color Theravadin countries in orange.
- Trace an arrow to show how Theravada Buddhism spread from India.
- Trace an arrow of a different color to show how Mahayana Buddhism spread outside India.

JAINISM AND BUDDHISM
AXIS AGE
ANCIENT INDIA
MYTH OF LIBERATION

JAINISM AND BUDDHISM: COMMONALITIES

- Originated in Ancient India in 6th c. BCE
- Common vocab: *Samsara, karma, maya, moksha*
- Sprang from Kshatriya caste
- Denied saving efficacy of Vedas + rituals
- Challenged Brahmin priesthood + castes
- Gods have to be reborn human to take path of enlightenment
- Women can be nuns

DIFFERENCES

- *JAINISM*
  - Extreme ascetic path
  - *Jiva* (True Self) [weighed down by *karma* and matter]
  - *Jiva* can save itself by discovering its own perfect, unchanging nature (No mergence in Brahman as in Hinduism)
  - Small group
  - Until end of 20th c. no missionary outreach
- *BUDDHISM*
  - Middle Way (moderation)
  - No self (*anatman*)
  - Made converts outside India
  - First missionary = Emperor Ashoka (Mauryan Empire)

CENTRAL PROBLEM OF INDIAN RELIGIOUS LIFE

How to find release from karma, escape the continual round of rebirth (samsara) entailed by it?

ORIGIN OF THE NAME JAIN

- Jain = one who follows a *jina*
- *Jina* = conqueror in Sanskrit
- Has been victorious over the obstacles to liberation
- Extreme ascetic path leading to liberation
- Liberation is the result of personal effort

NATAPUTTA VARDHAMANA: REMEMBER HIM AS MAHAVIRA (GREAT HERO)

- Was a contemporary of the Buddha
- Died in ca. 527 BCE
- Prince of a *kshatriya* clan

- At age thirty, started wandering as spiritual seeker
- Meditated and wandered naked
- Practiced strict *ahimsa* (non injury)
- Reached liberation after twelve years of meditation, silence, and extreme fasting

MAHAVIRA (GREAT HERO)

- Taught the path; followers came from all castes
- Left no writings (taught there is no creator god)
- Considered as the last of twenty-four savior beings called *Tirthankaras*
- *Tirthankaras* (ford-makers) are those who escaped the cycle of rebirth

JAIN TEXTS

- Agamas (= tradition)
- Contain the teachings of sages + sermons of Mahavira

JINAS AND TIRTHANKARAS

- *Jinas* = victors
- The *tirthankaras* = *jinas* who serve as models for others
- They are not gods
- Gods have to work out their own salvation
- Appear at each downward cycle to regenerate the world

BUDDHISM, A NONTHEISTIC RELIGION. WHY?

- No personal God
- No creator God
- Siddhartha Gautama = human being who attained full enlightenment through meditation
  - Originally Hindu prince [Kshatriya]
  - First took *sannyasin* + Jain path

INFLUENCE OF INDIAN THOUGHT

- Did founder of Buddhism intend to begin new religion? YES!
- Early Buddhist literature shows that Buddhism rejected
  - Vedas
  - Vedic practice
  - Vedic reliance on priests
  - caste system
  - gender and social limitations
  - belief in any permanent spiritual reality

ELEMENTS OF INDIAN THOUGHT ARE STILL VISIBLE

- Notion *of ahimsa* (non harm)
- Notion of rebirth called *samsara*
- Notion of *karma* (actions and consequences of actions)

- *Moksha* = nirvana in Buddhism
- *Nirvana* = release from suffering and rebirth that brings inner peace

FOUR NOBLE TRUTHS IN DEER PARK OF SARNATH

- Life is suffering
- Cause of suffering is desire
  - This desire makes us believe that there is something permanent and unchanging in life
- There is release from suffering
- Way to find release is to follow Eightfold Path

EIGHTFOLD PATH

- Right view
- Right aim
- Right speech
- Right action
- Right living
- Right effort
- Right mindfulness
- Right concentration

HOW COULD THE BUDDHA MAKE THE CLAIM THAT HIS METHOD WAS SUPERIOR TO OTHER PATHS HE HAD EXPERIENCED HIMSELF?

HINDU BELIEFS AND THE BUDDHA

- For Hindus, all living beings possess an immutable Self or soul (*Atman*)
- For the Buddha there is no self or no permanent identity (*Anatman*)

ANATMAN, REVOLUTIONARY CONCEPT

- No separate, permanent, or immortal self
- Human being = a composite of constantly changing states of being or *skandhas* in Sanskrit

SKANDHAS ("AGGREGATES")

- Body
- Perception
- Feelings
- Predispositions, generated by past existences, also called CONSCIOUSNESS, awareness of past mental events
- Reasoning

WHAT HOLDS THESE FORCES TOGETHER?

- Karma accounts for the seeming permanence of *skandhas*
- *KARMA* = state of mind of a person performing the action
- When the hold of the law of karma disappears, the forces will dissipate
- What is left is emptiness (*SHUNYATA*) = there is nothing to cling to (no permanent essence)

WITH NO PERMANENT SOUL, HOW CAN REBIRTH BE POSSIBLE?

- Elements of personality that make up an individual can recombine and continue from one lifetime to another
- In rebirth NO soul is transferred, only the karma-laden character structure of the previous life

GODS

- = humans
- Imperfect and impermanent
- Finite, subject to death and rebirth

GOAL OF BUDDHIST PRACTICE

- To escape "the chain of causation" and attain nirvana
- Nirvana = literally, extinguishing of a flame from lack of fuel
- Nirvana = cessation of all conditioned thoughts, that is karma

NIRVANA

- = quietude of heart
- At death, one enters deathless, peaceful state that cannot be described
- Possible to reach nirvana during one's lifetime
- Once a person has reached nirvana, no matter his/her caste, rebirth is finished

EARLY DEVELOPMENT OF BUDDHISM

- First two centuries after death of the Buddha, Buddhist monks spread across the Gangetic plain
- Several conferences of monks on monastic orthopraxy

TRIPITAKA ("THREE BASKETS")

- First Buddhist scriptures
- Written down in India about 80 BCE on palm leaves and stored in baskets
- Divided according to their subject into three groups (monastic guidelines, Buddha's teachings and sermons, nature of existence)
- In Pali language

KING ASHOKA OF THE MAURYAN EMPIRE

- In northern India
- 3rd c. BCE
- Renounced violence after conquering most of India and converted to Buddhism
- Condemned the slaughter of animals
- Sent missionaries beyond India, in particular Sri Lanka
- Helped Buddhism to spread from Afghanistan to the Bay of Bengal

AFTER THE DEATH OF ASHOKA

- Theravada Buddhism spread throughout Southeast Asia
- Among urban people and merchants
- 1st c. BCE, among Buddhist teachers, a new movement arose: Mahayana Buddhism

INTERNAL DIVISIONS

1. Theravada (Way of Elders)
2. Mahayana (= large vehicle)

THERAVADA BUDDHISM (WAY OF THE ELDERS) [SPREAD BY ASHOKA]

- Called Hinayana by Mahayana Buddhists (*Hina* = small) = derogatory term
- Each person must row himself/herself to the opposite shore in a small raft that holds only that person
- Emphasis on monkhood
- Model: Buddha (as teacher). He had no external assistance

MAIN CHARACTERISTICS

- Conservative and traditional
- Study of the Tripitaka
- Focus on life of Buddha
- Favor imitating the Buddha in monasticism
- Buddha was above all a man, a teacher
- Salvation is through dedicated self-effort rather than through intervention of heavenly beings

GOAL OF A THERAVADIN BUDDHIST

- To become an *arhat* (= perfect being, worthy), a person who has reached nirvana through his/her own effort
- Glorification of the example of the monk or *bhikshu* (the monk imitates the way of the Buddha)

WHAT ABOUT LAY PEOPLE?

- Lay people build up merit, so in a later life they would have a better chance to become enlightened
- By supporting the monks through their offering
- Doing good works
- Spending some period of life in monastic discipline

WHAT ABOUT WOMEN?

- About 1,000 years ago orders of nuns disappeared in Theravadin countries
- Women are considered capable of reaching nirvana, but spiritual power was kept in the hands of monks
- The texts edited by monks became actively misogynist (women are hindrances to monks' spiritual development)
- There are now attempts to revive fully ordained orders of nuns

MAHAYANA BUDDHISM

- = large vehicle of raft (*Maha* = large; *yana* = means)
- In the basic Indian worldview
  - "River" = cycle of rebirth
  - Far bank of the river = liberation from the cycle
- A large vessel, with a pilot, carries many persons to liberation
- There is more than one way to be liberated
- Emphasis is NOT on personal liberation but on liberation for all
- Reached China through Central Asia on silk routes

MAHAYANA

- Mainly China, Korea, and Japan
- Nirvana is possibility for everyone (NOT just for monks)
- Buddha has almost divine character
- Emphasis on Buddha's wonders and miracles
- There are many other Buddhas, enlightened beings who can help each generation
- Everyone is potential Buddha

"LOTUS SUTRA" (100 BCE-100 CE?)

- Most popular
- Contains sayings of the Buddha
- Reveals that the historical Buddha is but a manifestation of the real Buddha
- Real Buddha = cosmic Buddha who wants to show compassion for all beings (KARUNA)
- Came to earth in the form of a man because he loved mankind and wished to help
- Buddha did not die

MAIN GOAL

- To become like the Buddha by seeking enlightenment for the sake of saving others
- We are not called just to individual liberation but to save all

THERAVADIN GOAL # MAHAYANIST GOAL

- In Theravada, goal is to become an *arhat* (= perfect being, worthy), a person who has reached nirvana
- In Mahayana, goal is to become a bodhisattva (= future Buddha)

THREE MAJOR CONCEPTS OF MAHAYANA BUDDHISM

- TRIKAYA = Three Aspects of the Buddha
- KARUNA = compassion
- SHUNYATA = emptiness
- Goal = become bodhisattva + salvation for all
- Main writing = Lotus Sutra

TRIKAYA OR THE MAHAYANA COSMOS

- = three Bodies or Aspects of the Buddha
  1. THE COSMIC BODY or nature of the Buddha
  2. THE HEAVENLY BODY of the Buddha (among them Bodhisattvas)
  3. THE EARTHLY MANIFESTATION OR BODY of the Buddha (historical Siddhartha Gautama)

PLACE AND NUMBERS TODAY

- About 350 million today (= number of Protestant Christians)
- Fourth largest world religion after Christianity, Islam, and Hinduism
- Over 98 percent live in Asia
- 62 percent are Mahayana followers (north of tropics)
- 38 percent are Theravada followers (south of tropics)

THE SPREAD OF MAHAYANA BUDDHISM INTO CHINA

- Entered China through Central Asia in 1st c. CE
- Silk Roads
- Gained prominence in 7th c. under Tang dynasty

CHINESE PEOPLE REJECTED THERAVADA BUDDHISM

- Why?
- Ancestor worship in China made essential the continuation of male heirs

WHY DID MAHAYANA BUDDHISM SPREAD SUCCESSFULLY IN CHINA, KOREA, AND TIBET?

- Allowed many pre-Buddhist beliefs and practices to survive
- Indigenous gods became heavenly Buddhas
- Recognizes that people find themselves at different stages of spiritual evolution

STUPAS

- Shrine in the shape of a dome reaching into the sky
- Contains sacred relics of Buddha or remains of famous monks
- Marks sacred sites
- Built by the laity as an act of special merit

USE OF SYMBOLS IN WORSHIP

- In early Buddhism no images
  - Eight-spoked wheel (derived from the noble Eightfold Path)
  - Umbrella = Buddha's authority
  - Lotus flower
  - Symbols of a footprint or an empty throne
- In 1st c. CE, use of images (Greek influence?)
  - Buddha meditating, standing, walking

CLICKER NOTES

MYTH OF LIBERATION

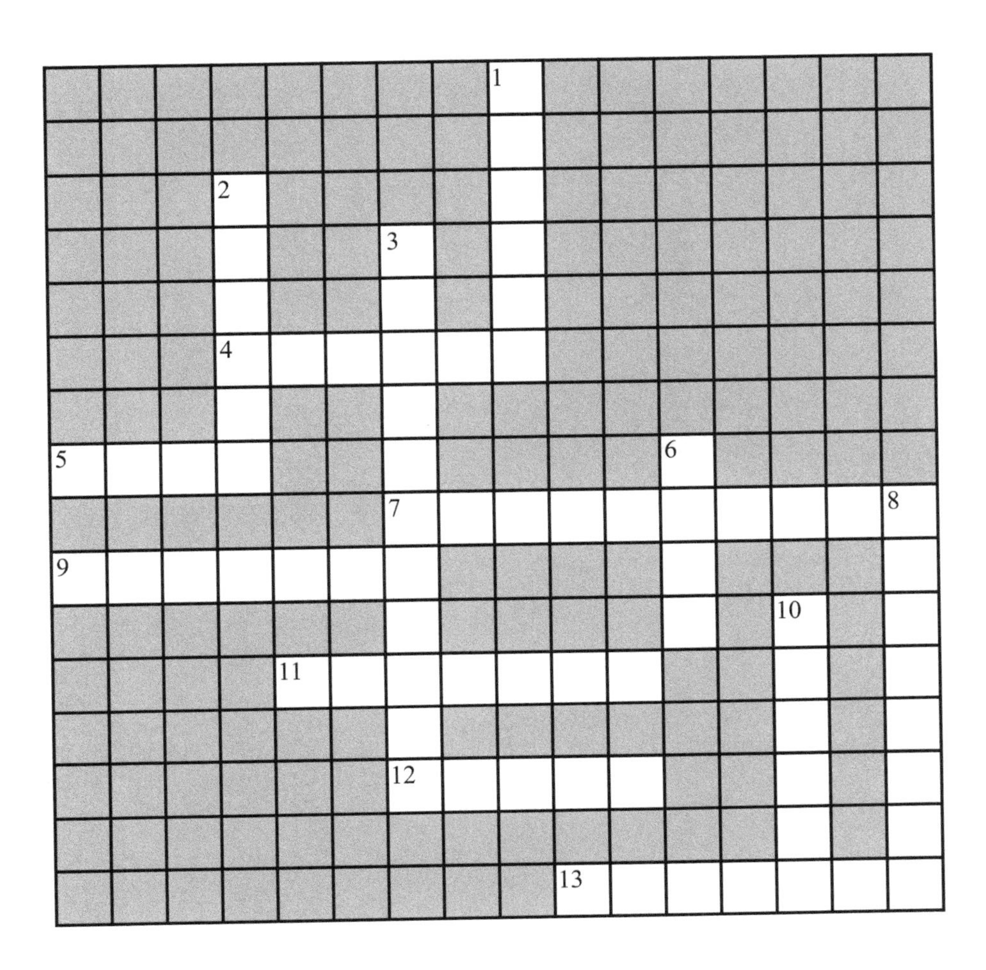

ACROSS

4. The totality of the Buddha's teachings
5. True self or soul in Jainism
7. Practice of extreme self-denial and renunciation of all worldly possessions (surest way to achieve moksha in Jainism)
9. No-self in Buddhism
11. Aggregate in Buddhism
12. True self in Hinduism
13. Moksha in Buddhism

DOWN

1. Siddhartha Gautama's clan
2. The Enlightened one (applied to Siddhartha Gautama)
3. "Crossing or ford-maker" (one of the twenty-four ideal human beings in Jainism, who taught the way to escape samsara)
6. Conqueror, victor over samsara
8. "Great Hero," or contemporary of the Buddha and founder of the Jain path
10. Non-violence

MAIN BRANCHES OF BUDDHISM

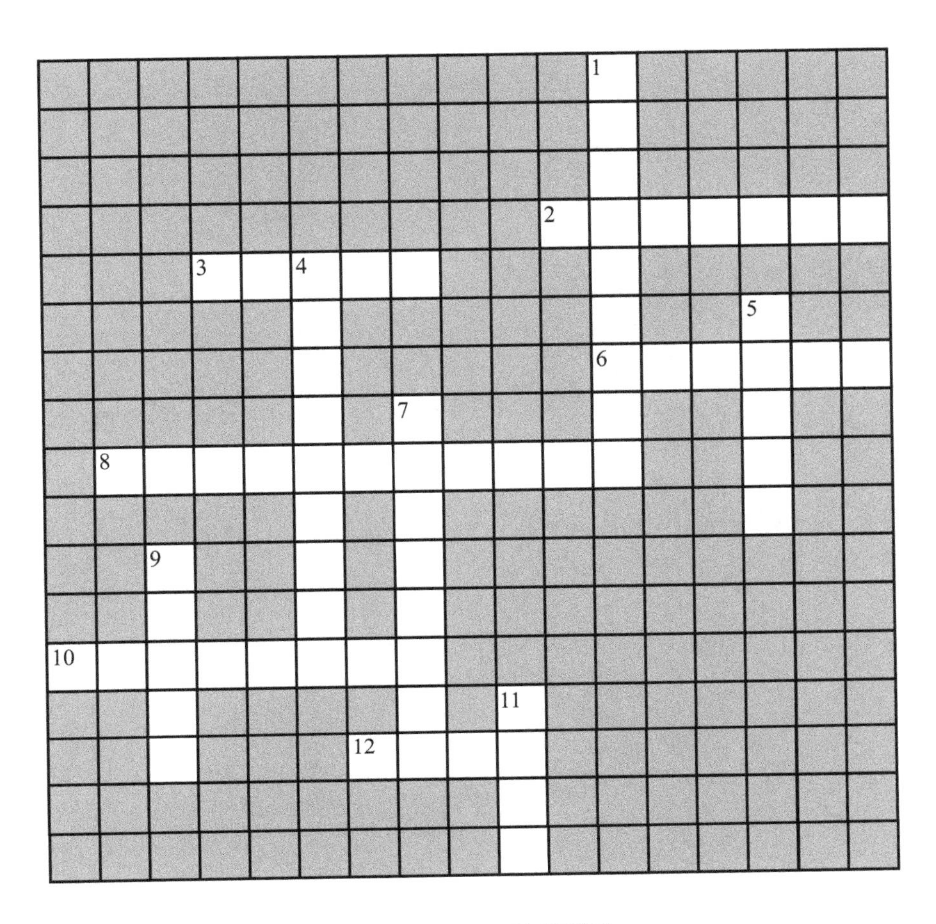

ACROSS

2. Three aspects of the Buddha (important concept of Mahayana Buddhism)
3. Sacred text that records the words of the Buddha
6. Famous king of the Mauryan empire in India who converted to Buddhism and sent missionaries to Sri Lanka, Afghanistan, and Ancient Greece (3rd c. BCE)
8. "Enlightenment being" or in Mahayana Buddhism, a person who has postponed his/her entrance into nirvana to help others in their quest for nirvana
10. Mahayana concept of emptiness (i.e. the universe is empty of permanent reality)
12. The first Buddhist scriptures were written in this language

DOWN

1. "The Way of the Elders" (this branch of Buddhism is prevalent in countries south of the tropics and stresses personal liberation from samsara).
4. The "Three Baskets," first Buddhist scriptures
5. This flower grows in swamps and symbolizes moksha
7. Large Vehicle (this branch of Buddhism is prevalent in China, Korea, and Japan). It promotes the idea of universal liberation through the help of bodhisattvas
9. Buddhist shrine in the shape of a dome
11. Buddhism and other religions spread through this famous road

LIFE OF THE BUDDHA

Write four ideas or facts that you learned from watching the video clips today.

1.

2.

3.

4.

Chapter 9

Daoism

Yin (Dark) and Yang (White) Symbols

Statue of Laozi in Quanzhou, China

Oracle Bones
(Divination)

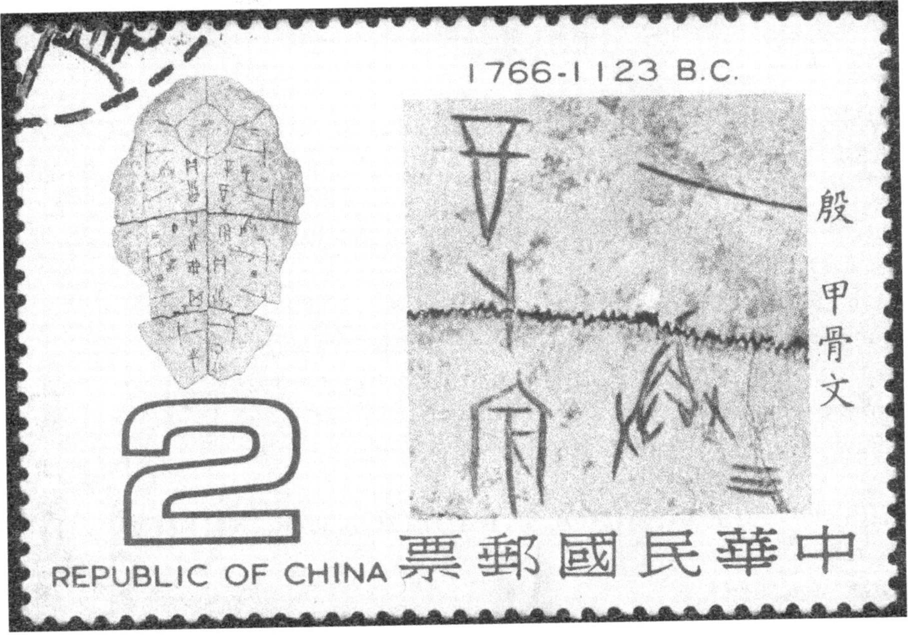

© YANGCHAO/Shutterstock.com

Stamp Printed in Taiwan Shows Inscriptions on Oracle Bones, c. 1976

© Zhao jian kang/Shutterstock.com

Terracota Warriors in Emperor Shi Huangdi's Tomb (Qin dynasty) (3rd c. BCE)

DAOISM

TWO MAJOR CONCEPTS TO UNDERSTAND CHINESE PHILOSOPHICAL AND RELIGIOUS TRADITIONS

- Harmony
- Virtue (*DE* in Chinese): moral excellence; goodness; righteousness

BEFORE AXIS AGE: SHANG DYNASTY

- First agriculture expanded along rivers (Yellow and Yangzi rivers)
- Shang dynasty (ca. 1600-1122 BCE) [first identifiable Chinese state]
- Developed bronze metallurgy (bronze objects found in graves)
- Used horse-drawn chariots (found in graves)

MAJOR FEATURES OF FOLK RELIGIOUS TRADITION DURING SHANG DYNASTY

- Veneration of ancestors
  - Sacred rituals are called li
- Veneration of deities and nature spirits
- Notions of *Dao, qi, yin,* and *yang*
- Belief in Shang Di, the Lord-on-High, Ruler of the Universe

SHANG DI

- Shang dynasty (1600-1122 BCE)
- Ruler of the Universe
- NOT a creator-god
- Supreme ancestor of the Chinese + ancestor of ruling Shang family
- Guarantor of moral order
- King = chief priest and diviner

EMERGENCE MYTH

- Not a god-creator. There is the Dao
- *Dao* (the Way) = mysterious source and ordering principle of the universe
- It is "zero" (that is, the pure potential)
- Source of gods
- Neither positive or negative
- No command; no judgment
- Experienced as female because it sustains and nurtures

HOW DOES THE DAO PRODUCE THE WORLD?

- Self-generation or gestation. The Dao grows itself into the universe
- The universe is the body of the Dao

STAGES OF COSMIC GESTATION

- *Dao* as primal chaos moved around until in its center was formed a "drop" of primordial breath

- Primordial breath or energy (*qi* or *ch'i*) divided itself
- *Yin* (femaleness) and *yang* (maleness) interacted and formed
- Heaven, earth, and humanity
- The Four Seasons
- The Five Directions (four directions plus the center) and the Five Elements [4 and 5 are the basic structure of universe]

QI (CH'I)

- Dao generated qi (ch'i)
- *Qi (ch'i)* = the One = primordial energy, the breath of life
- Each human being has this *qi*
- Daoist practice is based on how humans nurture *qi*

YIN AND YANG

- *Qi* developed two complementary patterns of energy or energy modes
- *Yin* = dark, heavy, obscure, passive, earth, death, and feminine
- *Yang* = light, warm, airy, active, life, heaven, and masculine
- The whole universe can be understood as *yin*, *yang*, or a combination of the two
- Both energies harmonized and formed human beings
- Human body = microcosm of world
- Cosmic connections can be found between body parts, seasons, planets, divinities, and the elements

WRITING

- In Mesopotamia and India, merchants pioneered the use of writing
- In China, the earliest known writing served the interests of the rulers
  - Early writings found on strips of bamboo, pieces of silk, and oracle bones
  - Written language included pictographs (conventional or stylized representations of an object); no phonetic or alphabetic component

MORE FEATURES OF FOLK TRADITION

- Belief in keeping harmony
- How can one keep harmony in the world?
  - Maintain worship of gods and ancestors
  - Divination
    - Reading cracks on heated bones and shells
    - Sticks or coins dropped on the ground to form trigrams and hexagrams
    - *I Ching* or *Yijing* ("Book of Changes")

AXIS AGE: PHILOSOPHICAL DAOISM

- Emerged between 6th and 3rd c. BCE under the Chou (Zhou) dynasty before first unification of China
- Time of major political and social chaos
- How to restore virtue (*DE*)?

THREE SCHOOLS OF PHILOSOPHY

- Legalist (or Authoritarian): human beings are basically selfish and materialistic. Clearly defined laws must be enforced to maintain order [Power comes from military might; children are raised to be loyal to the state]
- Confucian: virtue exists. Human beings need structure and rules in order to learn to love; family training
- Daoist school exalts nature over society. Virtue is understanding and conforming to the Dao

DEVELOPMENT OF DAOISM

- Started as a mystical philosophy of life (no focus on gods and no use of rituals)
- Developed into a popular religious system with many deities and many rituals

MANDATE OF HEAVEN (T'IEN MING)

- = the Dao of Heaven or the Order of Heaven
- More impersonal designation for concept of heavenly power
- Heaven
  - = deity, supreme reality
  - = divine order of universe
  - = self-existing moral law of virtue
- Heaven will choose a virtuous family to rule
- Special relationship based on merit, not birth

EMPEROR = SON OF HEAVEN

- Title held until beginning of 20th c.
- Heaven, earth, and the emperor linked together
- Ruler conducted rituals to maintain harmony between Earth and Heaven

DAOISM = AMBIGUOUS DESIGNATION

- Came to be to distinguish itself from Confucianism
  - Refers to philosophical tradition or *Dao jia* ("the Philosophy of the Dao")
    - Traced to legendary sage Laozi (Lao-tzu or Lao-tse)
    - Expressed in *Daodejing* ("The Classic of the Way and Its Power")
  - Religious tradition or *Dao jiao* (Teaching of the Dao) [1st-2nd c. CE]
    - Worship of deities through rituals
    - Human Laozi becomes Cosmic Laozi or Lord of Humanity (one of the Three Purities or Worthies of the Daoist pantheon of gods)

PHILOSOPHICAL DAOISM

- Scripture = *Daodejing* ("The Classic of the Way and Its Power") = political manual
- Author Laozi (or more than one author?)
- Systematized previous beliefs about Dao
- Main concern: how can individuals (rulers and subjects) experience Dao and let their lives flow in harmony with it?
- Book originally conceived as political manual

TO BE IN HARMONY WITH THE DAO. WHAT DOES THIS MEAN CONCRETELY?

- Live a simple, natural life
- Experience the unity of all things, rather than separation
- Accept and cooperate with things as they are
- Be receptive and quiet

A LIFE BASED ON WU WEI

- *Wu wei* = non action, inaction, non purposiveness
- = refraining from overly aggressive action
- = refraining from taking an intentional action contrary to natural flow of things
- Action without ego assertion

CONCRETELY IT MEANS

- Having no ambitions, no desires for fame and power
- Having no need to dominate others
- Leading a contemplative life + loving nature
- Virtue (*DE*) will be natural, NOT contrived (# Confucianism)

IN POLITICS, WU WEI = NON INTERFERENCE

- The ruler plays a passive role
- The ruler is hardly noticed
- The ruler takes the feelings and opinions of his people as his own
- If the government imposes strict laws, it will cause resentment and crime
- If it stresses morality, the people act artificially

RELIGIOUS DAOISM

- Developed in 1st-2nd c. CE
- Daodejing speaks of eternal Dao
- Seems to suggest that whoever is in harmony with the Dao lives forever

PURSUIT OF IMMORTALITY AND WELL-BEING

- If balance can be restored between yin and yang, humans can avoid death
- How?
  - Alchemy
  - Hygienic and dietary regime
  - Sorcery
  - Breath control techniques and physical exercises
  - Sexual practices involving the suppression of orgasm
  - Living a life of virtue
  - Worship in temples and petition to a deity or an immortal for assistance

DAOIST SOCIETIES

- Emerged in 2nd c. CE
- Had a political and religious dimension

- Responded to decline of Han dynasty (206 BCE-220 CE)
- Present unjust order had to be replaced by new Daoist imperium

RESPONSE TO BUDDHISM

- Buddhism entered China in 1st c. CE, and led to:
  - production of Daoist sacred books
  - opening of Daoist temples
  - groups of ascetics forming

MESSIANIC CHARACTER

- 1st c. CE, Laozi = immortal
- 2nd c. CE = deity = Lord of Humanity
- Laozi will manifest himself to save mankind from the Han rule
- Rebellions occurred, led by faith healers, advocating egalitarian ideas

THE WAY OF THE CELESTIAL MASTERS (1ST-2ND C. CE IN WESTERN CHINA)

- Founder: Zhang Daoling (healer)
- Group still exists today in Taiwan
- Secret society + strict hierarchy
- Goal: attain longevity through faith healing, meditative trance, and alchemy
- Accepted minorities and women in the ranks of parish leadership
- Priests and priestesses
  - Exorcized illness by prescribing confession of sin
  - Taught social ethics
  - Administered justice + collected family contributions

POLITICAL IMPACT

- Led revolts against government
- Thanks to this group, Daoism also received imperial recognition

RELIGIOUS DAOIST PANTHEON

- Organized on bureaucratic model of Han dynasty
- At the top *Dao*
- Then, *Qi* = primordial chaos or breath
- First category of deities (= emanations of the Dao)
  - Three Purities
  - Jade Emperor
  - Three Officials
- They hold courts in celestial paradises
- Second category of deities: immortals, ancestors, demons

THE THREE PURITIES = EMANATIONS OF THE DAO

- Lord of Heaven, Lord of Earth, and Lord of Humanity
- Lord of Humanity = Laozi

THE JADE EMPEROR

- Huang Di, the Jade Emperor identified with Shang Di, ruler of Heaven
- Jade Emperor = mythical emperor of ancient China
- Governed popular pantheon of regional gods

THE THREE OFFICIALS (OFFICIALS OF HEAVEN, EARTH, AND WATER)

- Ancient deities
- Keep records of human deeds on earth (stern)
- Control each person's life span and fate after death
- When a Daoist falls sick, priest would submit petitions to the Three Officials

CLICKER NOTES

THE SHANG DYNASTY

1. Where was the Shang kingdom located?

2. How did the Zhou dynasty portray the Shang dynasty?

3. What was found in Shang tombs?

4. What symbolizes the Shang power?

5. Where did the first military chariots come from?

6. Who practiced writing? What was the purpose of writing in China?

7. What is an oracle bone? Why are these oracle bones important for archaeologists today?

8. What was the role of the king?

1. Give the names of the main schools of Chinese philosophy and state their main differences.

   A.

   B.

   C.

2. Place these four Chinese dynasties in chronological order.

Han	•	•	1
Zhou	•	•	2
Shang	•	•	3
Qin	•	•	4

3. Was China unified under the Zhou dynasty?
   a. YES
   b. NO

4. What happened during the Qin dynasty?

5. What happened during the Han dynasty?

CHINESE PHILOSOPHY

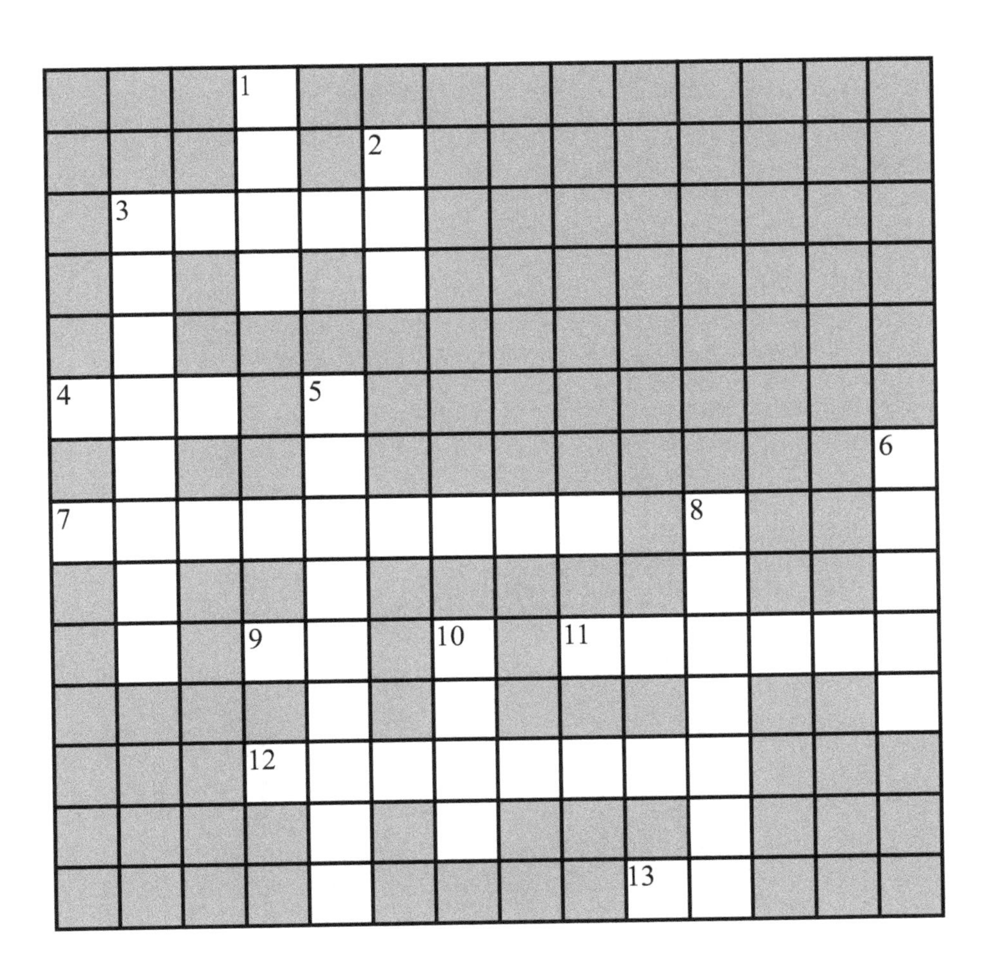

ACROSS

3. The Lord of Humanity (one of the Three Purities)and legendary founder of Daoism
4. The Way
7. Daoists view the body as a ________________ of the universe
9. Virtue in Chinese
11. ________________ bones were means of fore-telling the future during the Shang dynasty
12. Mandate of Heaven (or emperors' right to rule China, provided that they govern justly), a concept created by the Zhou to justify their rule
13. Breath of Life

DOWN

1. Famous dynasty of the 6th c. BCE
2. Receptive aspect of reality (associated with darkness, heaviness, femaleness, and earth)
3. School of Chinese philosophy advocating strict enforcement of laws by an authoritarian state (prevalent under the Qin dynasty)
5. The classic scripture of Daoism
6. Non action
8. Ruler of the universe and ancestor of the Chinese people
10. Active aspect of reality (associated with light, lightness, maleness, and heaven)

Chapter 10

Confucianism

Kong Fuzi or Confucius (Beijing, China)

ENTRANCE GATE OF CONFUCIUS GRAVEYARD (SHANDONG PROVINCE, CHINA)

Significance: Which dynasty in China gave full political support to Confucianism?

____

CONFUCIAN PHILOSOPHY

CONFUCIANISM IS A WESTERN TERM

- Term coined by Jesuit missionaries to China in 16th c.
- In China, Confucianism = *ru* tradition
- *Ru* = self-cultivation of moral virtue
- Dominant tradition in education and politics in East Asia until early 20th c.
- Confucianism is undergoing a current revival

HOW DO DAOISTS AND CONFUCIANS DIFFER FROM ONE ANOTHER?

- DAOISTS
  - Emphasized nature
  - Harmony of individuals with Dao
  - If humans let the Dao happen, virtue (*DE*) will come spontaneously
- CONFUCIANS
  - Emphasized better relationships
  - Harmony between cosmos order and social order
  - Humans make the Dao great by seeking virtue (*DE*)

THE FOUNDER: KONG FUZI (K'UNG FU-TZU)

- Great Master Kong (551-479 BCE)
- Confucius in Latin
- Poor aristocrat
- At age fifty, served as official in Lu
- His policies were rejected
- Became a wandering teacher

CONSERVATIVE?

- Did not question tradition, but accepted
  - The Lord on High (Shang Di)
  - Mandate of Heaven (*T'ien Ming*) [Heaven = power that rules universe. Determines what is right or wrong]
  - Ancestor worship
  - Spirits
  - Efficacy of rituals
- Emphasized the cultivation of moral virtues
- Believed political involvement can transform present world

REVOLUTIONARY?

- Challenged corrupt, autocratic leaders
- Argued for meritocracy
- Advocated education for all (boys, of course!)
- At age sixty-seven, he returned to Lu and spent the rest of his life teaching

CONFUCIAN TEXTS

- Four books
- Five classics [one of them is a book of divination = Yijing (I Ching) or Book of Changes]

THE FOUR BOOKS

- Analects (*Lun Yu*)
  - Sayings and conversations of Confucius
  - Compiled after his death
- Great Learning (Da Xue) = first text read by school boys about education of future *junzis*
- Doctrine of the Mean (Chung Yung). What is relationship between humanity and world order?
- Book of Mencius (3rd c. BCE)
  - Sayings of a principle disciple of Confucius
  - Systematized the teachings of Confucius

MENCIUS ("MASTER MENG," 371-289 BCE)

- Humans are by nature good
- Natural goodness must be cultivated
- Emphasized family obligations
- Rulers must provide for all citizens
- Just land distribution
- Right to revolt against an unjust ruler

CONFUCIAN TEACHINGS

- Humans are not primarily individuals. They are in relationships
- Five basic relationships:
  - parent and child
  - husband and wife
  - elder and younger brother
  - friend and friend
  - ruler and subject

BASIC PROBLEM OF HUMANITY

- Social chaos (Remember Zhou/Chou dynasty!)
- Chaos is caused by a breakdown of virtue
- Primary virtue = *ren* (benevolence) = will to seek the good of others
- Confucius aimed to restore social harmony
- Means of restoring harmony = education in virtue (*DE*)
- Virtuous or ideal person = gentleman (*junzi*)

INNER VIRTUES

- HAVE TO BE CULTIVATED WITHIN YOU
- *REN*, also written as jen = benevolence; sympathy for people's suffering = will to seek good of others
- *SHU* = reciprocity = do not do to others what you would not have them do to you
- *HSUEH* = self-correcting wisdom

OUTER VIRTUES

- SHOULD BE APPLIED IN YOUR RELATIONSHIPS WITH OTHERS
- *LI* (= propriety, good form)
  - Courtesy, respect
  - Right and proper order in family or social context
  - Religious rituals (ancestor veneration + worship of deities)
- *HSIAO* or *XIAO* = filial piety (devotion for parents)
- *CHENG-MING* = rectification of names = give people right titles

LEGALISM, DAOISM, AND CONFUCIANISM EVER PUT IN PRACTICE IN CHINA?

- LEGALISM
  - The Qin dynasty
  - Tradition of centralized imperial rule
  - Built roads to facilitate communication and movements of armies
  - Standardized laws, weights, writing
  - Adoption of a common script in writing
  - Adopted uniform coinage
  - Tomb of first emperor of China (Qin Shi Huangdi) buried with an entire army of life-size pottery figures
- CONFUCIANISM
  - Han dynasty
  - Tried to find a middle way between too much decentralization (Zhou dynasty) and too much centralization (Qin dynasty)
  - Brought Confucian-style government to Vietnam and Korea

HAN DYNASTY (206 BCE–220 CE) ADOPTED CONFUCIANISM

- Emperor sacrificed at Confucius's grave (became national shrine)
- Temples of Culture (*wen miao*)
  - *WEN* = culture, cultural refinement
  - Established in each province
  - Included images of Confucius
  - Sacrifices to Confucius
- Birthday of Confucius = holiday
- Four books + five classics = core curriculum
- Civil service exam for officials

CULT OF CONFUCIUS

- Confucius's descendants were ennobled
- Miracle stories about Confucius circulated
  - He could predict future
  - He was a god
  - He could perform miracles

WHAT ABOUT WOMEN?

- Pan Chao (Ban Zhao) = female Confucian scholar
- Argued that education should be available for girls
- But virtue most important for women is devotion to parents, husband, in-laws, and sons

PAN CHAO'S ARGUMENT

- The quality of yang is rigidity. Consequently men should rule
- The quality of yin is yielding. Women should yield and obey

OTHER MAJOR ACHIEVEMENTS OF HAN DYNASTY

- Advanced sericulture techniques (silk)
- Invented paper

CONFUCIANISM SPREAD BEYOND CHINA

- Japan
  - Confucianism enters in 7th c.
  - Becomes especially important in Tokugawa period (17th-19th c.)
- Korea + Vietnam

CONFUCIANISM IN CHINA TODAY

- 1980s—Confucius was rehabilitated
- His ancestral house rebuilt
- Confucian classics are studied

DAOISM AGAINST EMPEROR

- 2nd c. CE, Yellow Turbans conquered Yellow River in Eastern China
- Unequal distribution of lands + epidemics
- Yin and yang were no longer in balance in Heaven and on Earth
- A new "Yellow Heaven" mandate was to replace the Han dynasty mandate
- More militant than "The Way of the Celestial Masters" = emphasized violence and the end of the world

DAOISM SUPPORTED BY EMPERORS

- Thanks to the "Way of Celestial Masters," Daoism (rituals + medicine) received imperial recognition
- During Tang dynasty, which reunited China in 7th c. CE, candidates for the civil service were examined in Daoist + Confucian texts
- Tang imperial family claimed descent from Laozi

MAJOR DISCOVERY

- In 7th c. CE (Tang dynasty)
- Daoist priests discover gunpowder

IN COMPETITION WITH BUDDHISM

- Buddhism entered China through silk roads
- Very popular among nomads who later settled down
- Buddhists brought sugar, chairs + built temples (provided free education and lent money)
- Buddhism could give emperors legitimacy (emperors = enlightened beings who can bring others to enlightenment)

CONCLUSION

- Buddhism, Daoism, and Confucianism competed for state and popular resources
- At the popular level = religious pluralism

CLICKER NOTES

TEST YOUR CHINESE

1. Wu Wei •	• a. Empathy
2. Li •	• b. Virtue
3. Ren •	• c. Propriety
4. Dao •	• d. Breath of life
5. Qi •	• e. Male pattern of energy
6. Yang •	• f. The Way
7. De •	• g. Non-action
8. Junzi •	• h. Gentleman, aristocrat

Please format your answers as (1, a) below

CONFUCIAN ETHICS

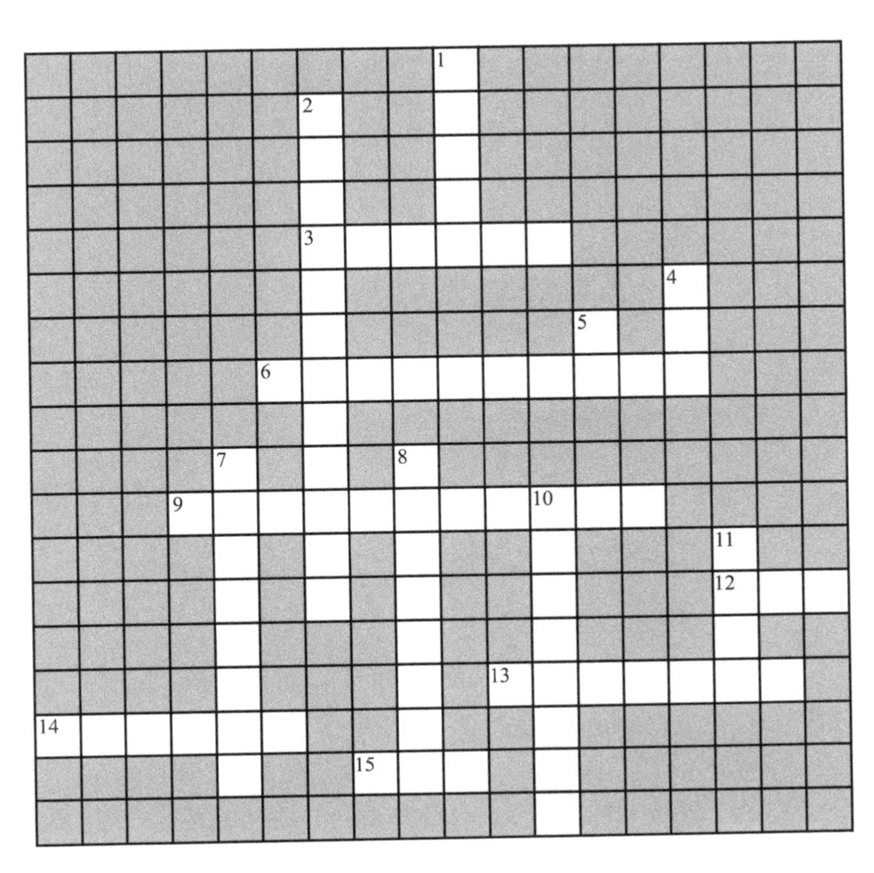

ACROSS

3. __________ piety (Hsiao) is the quality in children of showing respect and care for their parents and all ancestors
6. Foretelling the future
9. Government or institution in which individuals are awarded office based on their education and talents rather than on their birth or social status
12. Chinese dynasty that promoted Confucian texts and education (also time of deification of Laozi and Confucius)
13. Chinese philosopher (d. 289 BCE), who argued that rulers should provide for the common people
14. Naturalistic Chinese philosophy calling on people to live in harmony with nature
15. Chinese dynasty that favored authoritarian rule and outlawed Confucian texts

DOWN

1. "Noble person, aristocrat, gentleman" (the refined human ideal in Confucianism)
2. A system of thought in China based on proper human relationships and family respect
4. Benevolence, empathy
5. Propriety, ritual, etiquette
7. Authoritarian school of philosophy
8. Chinese name of Confucius
10. Book containing the sayings attributed to Confucius
11. Confucius lived at the time of the __________ dynasty

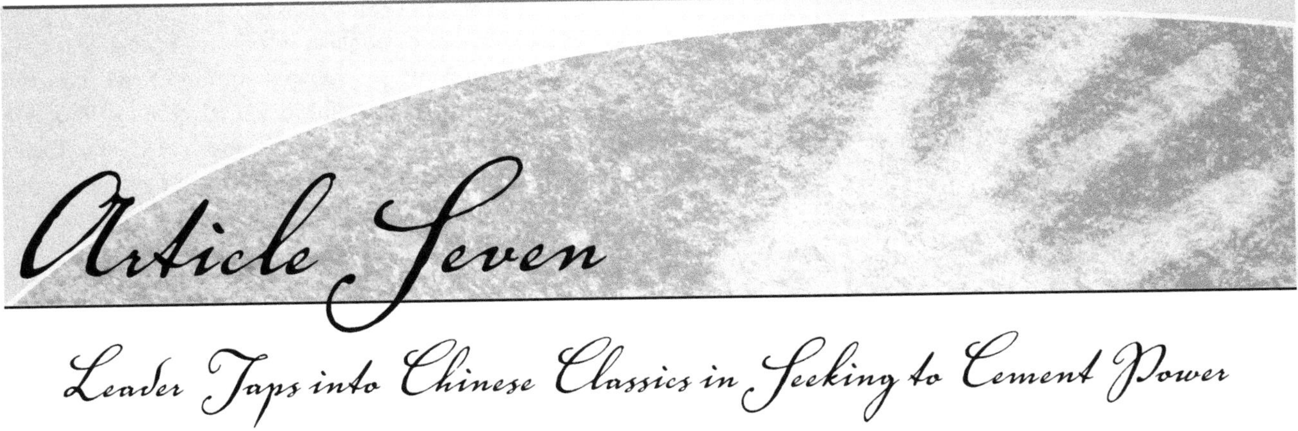

Article Seven

Leader Taps into Chinese Classics in Seeking to Cement Power

BY CHRIS BUCKLEY

HONG KONG — When China's leader, Xi Jinping, recently warned officials to ward off the temptations of corruption and Western ideas of democracy, he cited Han Fei, a Chinese nobleman renowned for his stark advocacy of autocratic rule more than 2,200 years ago.

"When those who uphold the law are strong, the state is strong," Mr. Xi said, quoting advice that Han Fei offered monarchs attempting to tame disorder. "When they are weak, the state is weak."

Seeking to decipher Mr. Xi, who rarely gives interviews or off-the-cuff comments, China watchers have focused on whether he has the traits of a new Mao, the ruthless revolutionary, or a new Deng Xiaoping, the economic reformer. But an overlooked key to his boldly authoritarian agenda can be found in his many admiring references to Chinese sages and statesmen from millenniums past.

Mr. Xi declared his reverence for the past last month at a forum marking 2,564 years since Confucius' birth. Ancient tradition "can offer beneficial insights for governance and wise rule," Mr. Xi said in the Great Hall of the People in Beijing, where leaders hold congresses and legislative sessions, according to Xinhua, the official news agency.

"This is about finding some kind of traditionalist basis of legitimacy for the regime," said Sam Crane, a professor at Williams College in Massachusetts who studies ancient Chinese thought and its contemporary uses. "It says, 'We don't need Western models.' Ultimately, it is all filtered through the exigencies of maintaining party power."

Since Mr. Xi became China's Communist Party leader nearly two years ago, he has pursued unyielding policies against dissidents, ethnic minority unrest, corrupt officials and foreign rivals in territorial disputes. The party has also rejected demands for democratic elections for Hong Kong's leader, and condemned two weeks of pro-democracy street protests in the city as lawless defiance inspired by foreign enemies of China.

At the same time, Mr. Xi has vowed to overhaul the economy, promote social equality, and build a fairer, cleaner legal system, which will be the main topic at a meeting of the party's Central Committee this month.

China's modern leaders have often sought to justify their policies by bowing to their Communist forebears, and so has Mr. Xi. But he has reached much farther back than his predecessors into a rich trove of ancient statecraft for vindication and guidance. He portrays his policies as rooted in homegrown order and virtues that, by his estimate, go back 5,000 years.

In his campaign to discipline wayward and corrupt officials, Mr. Xi has invoked Mencius and other ancient thinkers, alongside Mao. Most often, he has embraced Confucius, the sage born around 551 B.C. who advocated a paternalistic hierarchy, to argue that the party should command obedience because it represents "core values" reaching back thousands of years.

"He who rules by virtue is like the North Star," he said at a meeting of officials last year, quoting Confucius. "It maintains its place, and the multitude of stars pay homage."

In November, Mr. Xi visited Qufu, Shandong Province, where Confucius was born, to "send a signal that we must vigorously promote China's traditional culture." He told scholars that while the West was suffering a "crisis of confidence," the Communist Party had been "the loyal inheritor and promoter of China's outstanding traditional culture."

Mr. Xi has also shown his familiarity with "Legalist" thinkers who more than 23 centuries ago argued that people should submit to clean, uncompromising order maintained by a strong ruler, much as Mr. Xi appears to see himself. He has quoted Han Fei, the most famous Legalist, whose hardheaded advice from the Warring States era made Machiavelli seem fainthearted. And at least twice as national leader, Mr. Xi has admiringly cited Shang Yang, a Legalist statesman whose harsh policies transformed the weak Qin kingdom into a feared empire.

Their influence on Mr. Xi is likely to become clearer when a meeting of party leaders starting in just over a week endorses his proposals for "rule of law." Quite unlike the Western liberal version, Mr. Xi's "rule of law" looks more like the "rule by law" advocated by the Legalists, said Orville Schell, director of the Center on U.S.-China Relations at Asia Society in New York.

Mr. Xi wants party power to be applied more equitably and cleanly, but he does not want law to circumscribe that power, said Mr. Schell. This has created an "enormous amount of misinterpretation in the West that thinks 'rule of law' is rule of law in a very Western Enlightenment sense of the term," he said.

Mr. Xi's reverence for Confucius is a far cry from Mao, who attacked Confucian traditions as feudal dregs, to be erased by revolutionary fervor. By reviving tradition, Mr. Xi is riding China's nostalgic zeitgeist. Its people have increasingly turned to pre-Communist values while they navigate giddying, contentious changes driven by expanding commerce and inequality.

More children undergo Confucian-inspired coming-of-age rites, wearing re-creations of ancient scholars' gowns. Some universities have turned graduation ceremonies into rituals inspired by tradition. Devotees join in elaborate ceremonies to honor Confucius.

With Mr. Xi, 61, likely to be China's top leader for a decade, officials have been emulating him, and propaganda outlets have exhorted people to imitate his reverence for the ancient past. In May, the overseas edition of the state-run newspaper People's Daily published a selection of 76 of Mr. Xi's quotes from Chinese ancients, most often Confucius and Mencius, but also relatively obscure works that suggest a deeper knowledge of the classics.

"When Xi is putting on a political performance, he uses Marxist-Leninist rhetoric and even Mao's words," said Kang Xiaoguang, a professor of public administration at Renmin University in Beijing. "But in his bones, what really influences him is not those things but intellectual resources from the traditional classics."

This restoration of tradition has been encouraged by the party, eager to inoculate citizens against Western liberal ideas, which are deemed a decadent recipe for chaos. The Ministry of Education authorized guidelines in March to strengthen instruction in China's "outstanding traditional culture," and the party propaganda department has said traditional values are part of "socialist core values."

"As China grows stronger, this force for restoring tradition will also grow stronger," said Yan Xuetong, director of the Institute of International Studies at Tsinghua University in Beijing and author of "Ancient Chinese Thought, Modern Chinese Power."

"Where can China's leaders find their ideas?" he said. "They can't possibly find them nowadays from Western liberal thought, and so the only source they can look to is ancient Chinese political thinking."

Where Mr. Xi absorbed his enthusiasm for the classics is not so clear. He entered adulthood during the Cultural Revolution, when ancient tradition was under assault. But Mr. Xi has said he always liked to read, including Chinese classics, even as a teenager sent to labor in the countryside. Professor Yan of Tsinghua, who also came of age in the Cultural Revolution, said Mao's campaigns against Confucius helped introduce those very ideas to the young.

Visiting a university in Beijing last month, Mr. Xi said he lamented proposals that could reduce mandatory study of Chinese classical literature in school. He said, "The classics should be set in students' minds so they become the genes of Chinese national culture."

After reading article 7, explain the relevance of ancient Chinese thought for the world today.

Chapter 11

The Ancient Mediterranean: Persia, Greece, and Rome

Map of the Roman Empire (2nd c. CE)

Part I: PERSIA

© Borna_Mirahmadian/Shutterstock.com

Persian Achaemenid Soldier (6th BCE?)

© Anton_Ivanov/Shutterstock.com

Tomb of Cyrus the Great (Pasargadae, Iran)

© Anton_Ivanov/Shutterstock.com

Zoroastrian Tower of Silence (Dakhma) in Yazd (Iran)

THE PERSIAN EMPIRE

CHANGE YOUR PERCEPTION ABOUT THE MEDITERRANEAN SEA

- Mediterranean Sea: a watery barrier separating three continents? NO
- Rather an internal sea biding people all around it
- Formed integrated cultural, economic, and political unit
- Easier to travel by sea than by land

ECOLOGY

- Same climatic zone: hot dry summers and mild rainy winters
- Land difficult to cultivate (susceptible to drought)
- Surrounded by either mountains or desert
- Seaborne transportation easier and faster than land transportation
- Agriculturalists had to diversify economy (trading, fishing, and manufacturing)

CHRONOLOGY

- By 6th c. BCE dominated by city-states (Greece) + expanding Persian empire
- By 1st c. BCE-1st c. CE Rome is dominant
- By 6th c. CE decline of cities + Barbarian invasions from the north
  - Roman Empire collapsed in the West
  - But survived in the East (Byzantium/Constantinople)

MAIN CHARACTERISTICS

- Formation of vast imperial multiethnic states
- Emergence and consolidation of new monotheistic faiths (Zoroastrianism, Rabbinical Judaism, and Christianity)

THE ACHAEMENID EMPIRE OF PERSIA [FIRST GREAT EMPIRE]

- First Achaemenid king = Cyrus the Great
- Mentioned in Bible (defeated Babylonians and allowed Jews to return to Palestine to rebuild the second temple in 6th c. BCE)
- Demonstrated how it was possible to maintain vast multiethnic imperial states over a long period of time

MODEL FOR LATER EMPIRES

- No pillage + no enslavement + partnership in administering empire
- Subjected people's administration and military structures remained in place
- People could retain own religions and cultures
- Trade and cultural exchanges fostered
- Nomadic people contained (autonomy in exchange of military help to expand/defend empire)

DARIUS (521-486 BCE) [CAPITAL CITY: PERSEPOLIS]

- Chose strong central rule
- Divided empire into satrapies (provinces), all ruled by governors who were Persians (related by birth or marriage to ruling family)
- However, local administrators under them were members of the ethnic group they were supposed to rule
- Parallel military force (imperial spies) for surprise inspections

ORGANIZATION + COMMUNICATION

- Standardized taxes + measurements
- Codified laws (Mesopotamian model)
- Maintained roads (paved with stones) [Persian Royal Road]
- Organized mounted postal service
- Built *qanat* (underground canals) to irrigate land

PERSIANS AND NOMADS

- Fought against Scythian tribes
- Gave local autonomy to nomads
- Used them in armies
- Borrowed steppe compound bows

ZOROASTRIANISM

- Place of origin: Iran
- 10th c. BCE? 6th c. BCE?
- Monotheistic? Dualistic?
- An archangel sent by AHURA MAZDA (the Wise God) appeared to Zarathustra (Zoroaster)
- Before Zoroastrianism, Iran was polytheistic/animistic
- Islam entered Iran in 7th c. CE

SIGNIFICANCE

- Zoroastrianism influenced
  - Judaism
  - Christianity
  - Islam
- All four emerged in Middle East
- BUT Judaism, Christianity, and Islam
  - Claim descent from Abraham
  - God called Abraham, who
    - renounced other gods
    - lived by faith in the one true God
- This does not apply to Zoroastrianism
- HOWEVER, Zoroastrianism had impact on eschatology of all three
  - Heaven/Hell
  - Satan
  - Judgement Day

COMMONALITIES

- All four religions
  - are monotheistic
  - believe in one all-powerful personal deity
  - struggle between good and evil

GOD IS

- One
- Holy, righteous, immortal
- Creator
- Omniscient, omnipotent
- He has no equals
- Transcendent
- Personal
- Beyond gender

HUMANITY IS CENTRAL TO CREATION

- Superior to other creatures
- Special relationship with God
- Special responsibility

SPECIAL HUMAN STATUS

- Reason
- Freedom
  - Moral choice
  - Capacity for good and evil

HUMANITY AND CREATOR

- Creator has revealed the good path
- The path is revealed in scriptures
  - *Avesta* (= instruction, law) is the sacred Zoroastrian liturgical text written in Avestan, ancient Iranian language
  - Book assembled in ca. 3rd. CE

TWO ETERNAL SPIRITS

- Spenta Mainyu (a good spirit)
- Angra Mainyu (an evil spirit, "the destructive spirit") = also called Satan
- Both spirits emanate from the one God, Ahura Mazda (= fatherly figure, expresses his will through Spenta Mainyu and higher spirits named Good Thought, Piety, Prosperity)
- Both good and evil spirits exist in balance and are necessary for life

ONLY ONE LIFE TO LIVE

- Zoroastrianism emphasizes human accountability
  - Good deeds are rewarded in paradise

- Evil deeds are punished in hell
- On the fourth day the soul is judged
  - Deeds are weighed on scale

ZOROASTRIAN ESCHATOLOGY

- At end of time, Ahura Mazda wipes away all evil
- Apocalyptic battle between good and evil
  - Resurrection of dead to fight for good or evil
- Angra Mainyu will be destroyed
  - New age will begin
  - No more evil, death, or disease

HOW TO BE ONE WITH GOD?

- Good thought
- Good word
- Good deed
- Worship God
  - Sacred fire in Fire Temples symbolizes God's presence

ZOROASTRIAN FUNERARY RITES

- Corpses are impure
  - Corpses cannot be buried or burned
  - This would defile earth or fire
- Corpses are exposed to carrion birds
  - Corpse is bound in roofless *dakhma* (tower of silence)
  - Corpse eaten by vultures

LINEAR TIME

- Zoroastrianism = first linear concept of time
- Existence moves from its creation by God to a final culmination
- The world is real, not illusion
  - Proving ground for humans

WHAT HAPPENED NEXT IN THE HISTORY OF PERSIA?

- Alexander the Great conquered Persia (4th c. BCE) and used same techniques as the Achaemenids to rule empire
- Seleucids (Seleucus Nikator was one of Alexander's generals)
- Parthians (originally Central Asian nomads)
- Sasanians/Sassanid who institutionalized Zoroastrianism as the state religion of Persia in 3rd c. CE
- Islam arose in the 7th c. in Arabia and took place of Zoroastrianism as state religion in Iran after conquest

CLICKER NOTES

THE PERSIAN EMPIRE

What did you learn today about __________ ?

1. CYRUS:

2. DARIUS:

3. XERXES:

4. PERSEPOLIS:

PERSIA

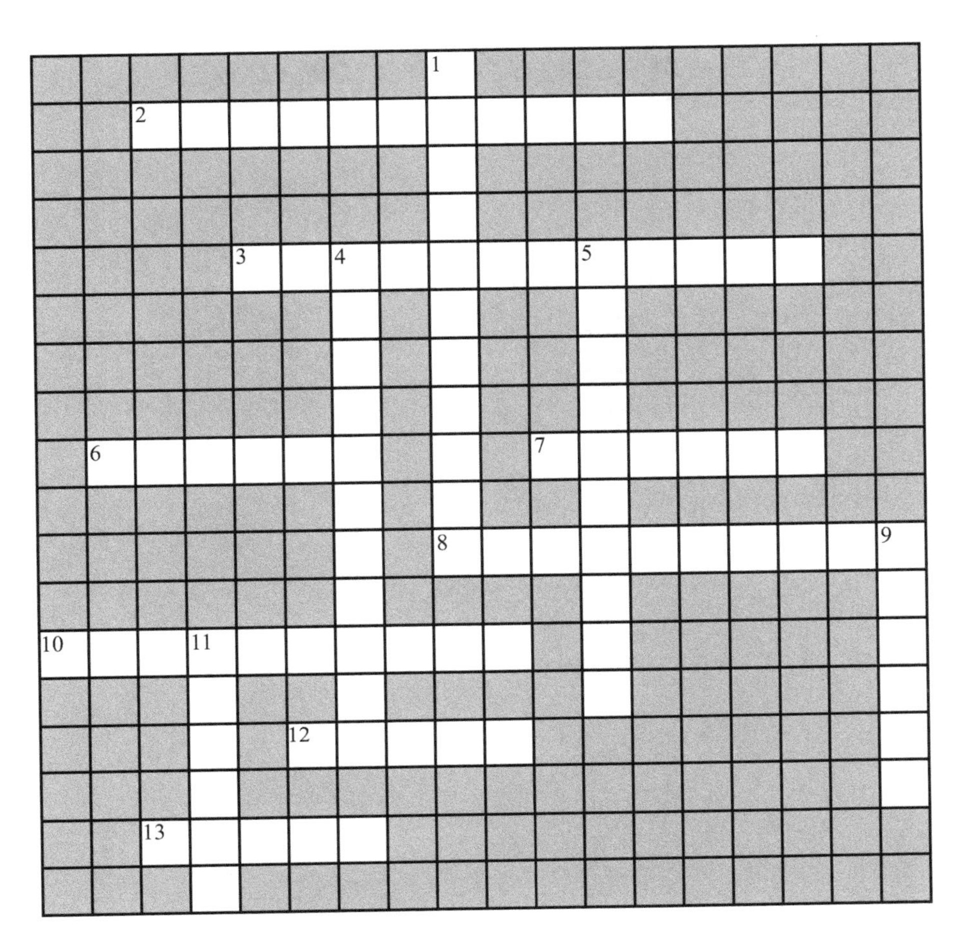

ACROSS

2. Evil spirit or Satan
3. Good spirit
6. Sacred book of Zoroastrianism (collection of hymns and laws)
7. Famous Persian king who founded a new royal residence in Persepolis and reorganized the Royal Road through Persia and Anatolia.
8. Famous Persian dynasty responsible for creating the very first multi-ethnic imperial state
10. Capital city of the Persian Empire erected by Darius the Great and Xerxes
12. First Achaemenid Persian king who respected the institutions and beliefs of subject people
13. Underground canal system in Persia

DOWN

1. An archangel appeared to ______________ and declared that there was only one god, the Wise God
4. Discourse about the "last things" (end of the world, heaven, hell, and judgment)
5. The Wise God in Zoroastrianism
9. Tower of silence
11. A governor with both civil and military powers in the Persian Empire

Part 2: GREECE

© Sorin Colac/Shutterstock.com

The Parthenon in Athens (Attica)

© Lefteris Papaulakis/Shutterstock.com

Aristotle on a Greek Stamp Dedicated to his 2300th Death Anniversary (ca. 1978)

© abxyz/Shutterstock.com

Pericles (5th c. BCE)

GREEK CIVILIZATION

ANCIENT GREECE [NOTION OF CITIZENSHIP]

- Politics = polis (Greek for city-state)
- Citizenship = "having a share in the public life of the polis"
- Citizenship did not imply equality
- Citizenship tied to military service (only men with resources could afford the military equipment) + ethnic identity

EXPANSION OF CITIZENSHIP

- Why?
  - Commercial growth enlarged taxpaying population
  - Military innovations reduced expense required for fighting
- Military innovation: move from expensive chariot fighting/cavalry to hoplite warfare
  - *Hoplite* = heavily armed foot soldier fighting in dense mobile formations (*phalanx*)
  - *Phalanx* = group of heavily armed infantry with large shields that overlap with the shield of the adjacent men

SPARTA IN SOUTH-EASTERN PELOPONNESUS

- Military oligarchy but not a military dictatorship
- All adult male = citizens (except for helots = subjected neighboring people who tilled land)
- Citizens were eligible to attend public assembly [passed laws]
- Execution of laws entrusted to two hereditary kings, acting with a council of twenty-eight members (chosen by public assembly)

ATHENS IN ATTICA

- Did not impose order by military means
- Not everyone was citizen of state (foreigners, slaves, and women had no direct voice in the government)
- 6th c. BCE, tensions grew between rich aristocratic landowners and owners of smaller plots
- Debts overwhelmed many and pushed them into slavery, which led to violence + expansion of citizenship

ARISTOCRAT SOLON (AXIS AGE = ENLARGED CITIZENSHIP)

- Forged compromise between all classes
- Forbade debt slavery + canceled all debts
- Opened councils of polis to any citizen wealthy enough to devote time to public affairs
- Later reformers paid salaries to office holders so that financial hardship would not exclude anyone for service
- Under Pericles (5th c. BCE), hundreds of officeholders came from common classes

STATUS OF WOMEN

- *SPARTA*
  - Girls and boys were raised the same way with an emphasis on physical training and horsemanship

- Often heads of family (because husbands in barracks)
- Could own property and do business without husband's consent
- *ATHENS*
  - Confined to house (ran the household, supervised slaves, and wove clothing for family)
  - Under protection of male members
  - After menopause freer (nurses, midwives)

HOMOSEXUALITY

- Acceptable among upper-class men
- A homosexual relationship between a young man and his master/mentor was often part of the young man's apprenticeship
- Did not preclude marriage with a woman and fathering a family
- Homosexuality was acceptable in the army. Male couples often served together

GREEK CULTURE AND RELIGION

- Greeks were polytheistic
- Greeks rarely became irreligious
- Dramas, civic festivals, and Olympic games were all religious celebrations

OLYMPIC GAMES

- Every four years Greeks engaged in contests of strength (foot racing, long jump, boxing, wrestling, javelin tossing, and discus throwing)
- Games = religious celebrations. Athletes offered their *arete* (their best, virtue, excellence, rightness) to the gods
- Games helped create sense of collective Greek identity all around Mediterranean Sea

GREEK DRAMA

- Rose out of the worship of Dionysus, god of the grape harvest, wine, fertility
- Wealthy Athenians financed productions and judges awarded prizes for best tragedies
- Dramas based on mythological and historical themes to probe fundamental problems of morality

GREEK PHILOSOPHY

- **Before Axis Age** = mythical thinking for natural phenomena
- **After Axis Age** = applied reason to the physical world
  - Capricious gods do not manipulate nature
  - Underlying the seeming chaos of nature there are principles of order, general laws
  - These laws can be investigated through reason

SOCRATES

- What is the place of the individual in society?
  - Individuals (not gods) at the center of the universe
- Is it possible to arrive at universal standards of right and justice? Yes, through reason
- How can moral excellence be achieved?
  - People should regulate their behavior in accordance with universal values

- True education = shaping of character according to values through the active use of reason
- Dialectics (or logical discussion) is a way of questioning assumptions, dogmas

PLATO: THEORY OF IDEAS

- There exists a higher plane of reality = realm of ideas
- Realm of ideas = universal standards of beauty, goodness, absolute
- Good life = to live in accordance with these standards
- Only philosophers can leap from worldly particulars to an ideal world beyond space and time

PLATO: THE JUST STATE (THE REPUBLIC)

- Living an ethical life will be easier if state is rational and just
- The just state will not be founded on tradition (for inherited attitudes do not derive from rational standards)
- Plato did not advocate democracy (democracy gives power to most popular, not to most knowledgeable)
- Only philosophers can lead the state (but not the commoners who cannot think intelligently)
- Their rule should be absolute
- Philosophers would be chosen through a rigorous system of education

ARISTOTLE

- Alexander the Great's tutor
- Differed from Plato
- Did not believe that to comprehend reality one needed to escape into another world (World of Ideas)
- Universal ideas cannot be determined without examination of particular things
- Favored development of empirical sciences (physics, biology, zoology, botany)

GREEK SCIENCES

- Democritus suggested that all physical matter was composed of indivisible particles called atoms
- Hippocrates rejected supernatural causes and based medical practice on the understanding of human anatomy and physiology

MACEDONIANS

- Alexander the Great avoided steppes (even built walls against nomads)
- Greek colonists brought language, gods, architecture to India, Central Asia, Persia

IMPACT ON MONOTHEISTIC THOUGHT

- Platonism is a two-world philosophy
  - Believes in a higher world as the source of values
  - Believes in the soul's immortality
- It had an enormous effect on religious thought (Judaism, Christianity, and Islam)
- Aristotle (prime mover) (= God in monotheistic thought)

CLICKER NOTES

ANCIENT GREECE

Please write four ideas or facts that you learned from watching the video clips today. Concentrate on Greek inventions.

1.

2.

3.

4.

WHO ARE THESE GREEK GODS?

© Hein Nouwens/Shutterstock.com

© MaKars/Shutterstock.com

Picture 1. ______________________

Picture 2. ______________________

© Hein Nouwens/Shutterstock.com

Picture 3. ______________________

GREECE

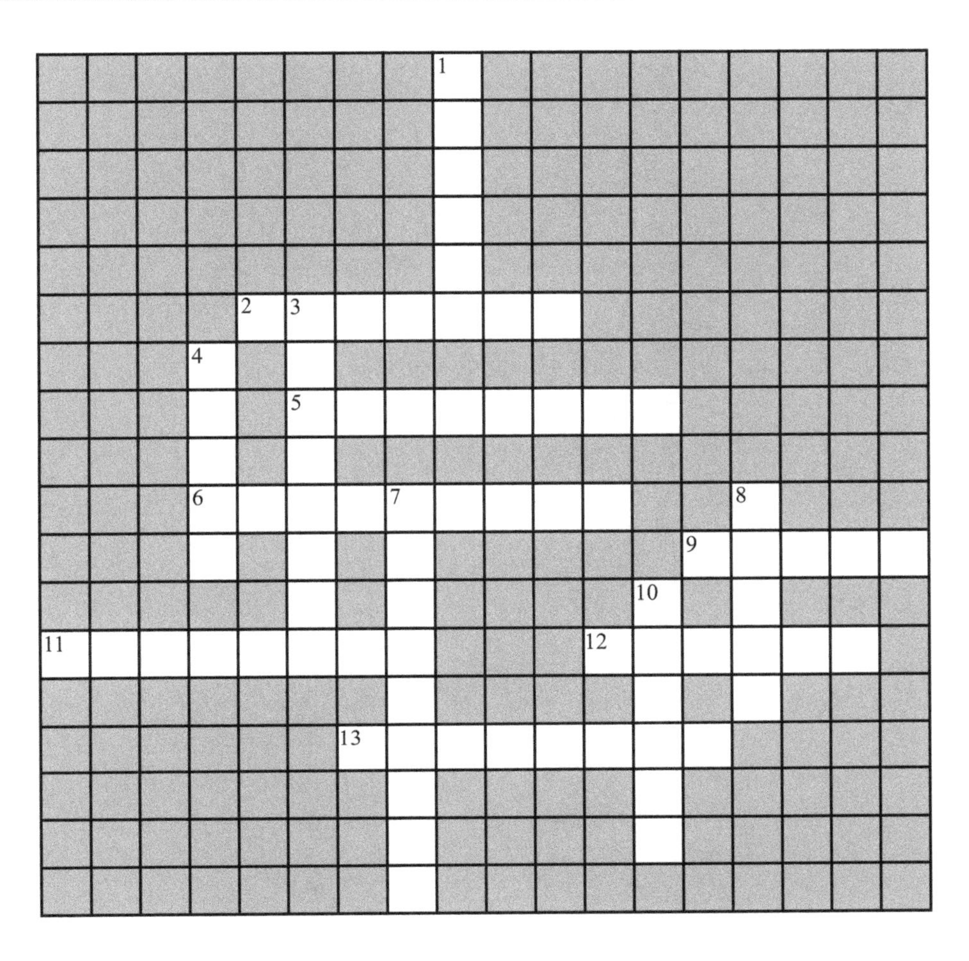

ACROSS

2. A rectangular formation of tightly massed infantry soldiers
5. Aristocratic leader who led the Athenian state to full participatory democracy for all male citizens
6. Rule by a select few
9. Aristocrat in Athens who forged a compromise between wealthy aristocrats and indebted landless common classes (6th c. BCE)
11. Greek god of wine, also known as Bacchus, in honor of whom plays were performed
12. The region surrounding the ancient city of Athens
13. Famous Athenian controversial philosopher who argued that one can arrive to universal standards of right and justice through reason

DOWN

1. This Greek city was a military oligarchy
3. Heavily armed infantry soldiers used in ancient Greece in a phalanx formation
4. Subjected neighboring people (serfs) who tilled the land or servants to the Spartan state
7. Great philosopher and scientist who favored the development of empirical sciences (biology, botany, zoology, physics, etc.)
8. Greek word for city-state, a politically sovereign urban center with adjacent agricultural land
10. Goddess of wisdom and war

Article Eight

When the Games Began: Olympic Archaeology

BY JOHN NOBLE WILFORD

Opening day of the ancient Greek games was a spectacle to behold, a celebration of the vigor and supercharged competitiveness that infused the creative spirit of one of antiquity's most transforming civilizations.

People by the thousands from every corner of the land swarmed the sacred grounds, where altars and columned temples stood in homage to their gods. They came from cities that were often bitter rivals but shared a religion, a language and an enthusiasm for organized athletics. There was no doubt in their minds that the games were as much a part of Greek culture as Homer, Plato or Euripides, and on a summer day at Olympia, perhaps more so.

At dawn, the opening procession of athletes began: runners and jumpers, discus and javelin throwers, boxers and wrestlers and charioteers, all young men, marching to the stadium and the hippodrome. They went from one altar to the next and past shrines to heroes of previous games. Finally, a trumpet sounded the beginning of the big event.

The exuberance and pageantry of the original Greek games -- even the spirit of community among rivals, however fleeting -- will be re-enacted in August at the next modern Olympic Games. They will be held in Athens, in the land where it all began.

A closer study of ancient texts, art and artifacts and deeper archaeological excavations are giving scholars new insights into the early games and just how integral athletics was to ancient Greek life. The games, said Dr. Stephen G. Miller, an archaeologist who is a classics professor at the University of California, Berkeley, "ran hand in hand with Greek cultural development."

For almost 12 centuries, starting as early as 776 B.C. at Olympia in the Peloponnesus, organized athletics were so popular that nothing was allowed to stand in the way. When it was time for the games, armies of rival cities usually laid down their weapons in a "sacred truce." In 480 B.C., while the Persians were torching Athens, there was no stopping the foremost games at Olympia.

In athletics, scholars are finding, the ancient Greeks expressed one of their defining attributes: the pursuit of excellence through public competition. The games were festivals of the Greekness that has echoed through the ages and reverberates in the core of Western culture.

"Of all the cultural legacies left by the ancient Greeks," Dr. Edith Hall of the University of Durham in England has written, "the three which have had the most obvious impact on modern Western life are athletics, democracy and drama."

As Dr. Hall noted in the Cambridge Illustrated History of Ancient Greece, all three involved performance in an adversarial atmosphere "in open-air public arenas in front of a large mass of often extremely noisy and critical spectators."

In these competitive exhibitions, she added, "success conferred the highest prestige, and failure brought personal disappointment and public ignominy."

Dr. Donald G. Kyle, a professor of ancient history at the University of Texas in Arlington, said that long before the Greeks, others engaged in competitive sports like running and boxing. Contemporaries of the Greeks in Egypt and Mesopotamia put on lavish entertainments at court, with acrobats and athletes performing, and also promoted some sports as part of military training. Dr. Kyle is writing a book on sport and spectacle in the ancient world.

But the Greeks, the historian said, took athletics out of the court and into the wider public, beyond the singular spectacles to regularly scheduled competitions. They spread their games as they colonized Sicily and southern Italy and Alexander the Great conquered Eastern lands, he said, "in the same way the British took cricket everywhere they went."

"The Greeks linked their games to recurring religious festivals," Dr. Kyle said, "and this regularized and institutionalized athletics."

In "Ancient Greek Athletics," a book being published next month by Yale University Press, Dr. Miller of Berkeley has sifted through literature, art and recent archaeology to compile a comprehensive history of sports in ancient Greece and their relationship with social and political life. Dr. Kyle called the book "very authoritative, really a momentous publication."

Dr. Miller dates the origin of Greek organized athletics to the beginning of the eighth century B.C. The Greeks were awakening from a "dark age" of several centuries and were energized by the arrival of the Dorians, people who probably came from the north. Archaeological finds suggest a sharp growth in population, prosperity and substantial architecture. Soon afterward, the first known Olympic games were held at the sanctuary of Zeus in Olympia -- not, as is sometimes supposed, on Mount Olympus, which is farther north.

In the same century, Homer's "Iliad" described footracing, wrestling, boxing, discus throwing and other contests at the funeral of a fallen Greek. But Homer, scholars think, was drawing on sports of his own time, not of the Trojan War five centuries earlier.

Olympia was the site of only one of the four major competitions, the Panhellenic games. The others were at Delphi, begun in 586; Isthmia, near Corinth, in 580; and Nemea, in 573. The contests were held every four years at Olympia and Delphi, every two at the other sites. Most cities also had their own local games.

Much of the new research draws heavily on texts of ancient writers, inscriptions on stadium walls and statue bases, and artifacts excavated from ruins at the sites of the contests, including the stones of starting lines and turning points for races. At Nemea, where he has excavated since 1973, Dr. Miller said he discovered what he feels sure was the stadium's locker room, the earliest known inner sanctum of athletes.

Archaeologists have also recovered the jumping weights and discuses of athletes and jars that held the olive oil they rubbed on their bodies, probably to warm up before exercise or a race.

"The largest single category of visual evidence is the vase painting," Dr. Miller said. The pentathlon, five competitions that a single winner had to excel in, seemed to be a favorite of vase painters.

Fans of the modern Olympics, Dr. Miller and other scholars said, would find striking differences at the original games. There were no team sports and no second-place prizes. Fouls were punished by flogging; vase paintings show judges with switches. The athletes, though considered amateurs, were allowed to accept cash and valuable gifts before and after competing. Women were prohibited from watching or taking part in the games, except as owners in the horse races.

In later years, though, some separate contests were staged for women in honor of Hera, the wife of Zeus. Dr. Jenifer Neils, an art historian at Case Western Reserve University who specializes in ancient Greek culture, said that unmarried girls ran a footrace wearing the Greek equivalent of a gym tunic that left the right breast bare, a foreshadowing, perhaps, of Janet Jackson's Super Bowl stunt, she suggested.

The most obvious difference in the early games was the nudity of the young men. The reasons are obscure. In one story, the practice started after a runner in an early contest lost his loincloth but continued on and won the race. Or it may be, as Dr. Neils said, that "in a body-conscious society, the robust nude male was the ideal form, another expression of Greek competitiveness."

In his book, Dr. Miller reconstructs the scene at one of the Panhellenic games in 300 B.C., the heyday of Greek organized athletics.

For days before the first races, crowds feasted on the meat of oxen roasted on altar fires, sacrifices overseen by a priest and accompanied by a flutist, a libation pourer and libation dancers. People pitched hundreds of tents across fields and thick smoke from campfires filled the air. Flies swarmed, speakers ranted and fights broke out.

There was something for everyone in the noisy throng: magicians and fortunetellers, poets reciting verse and sculptors displaying their works, and, in the account of one ancient writer, "countless lawyers perverting justice and not a few peddlers peddling whatever came to hand."

The young athletes, the pride of cities far and near, had been there for a month of supervised training. They aspired to the simple prize of the laurel garland. The word "athlete" is from an ancient Greek word that means "one who competes for a prize." Beyond that, however, winners could expect parades back home, statues and poems in their honor, a lifetime stipend and enduring fame.

"Sport for sport's sake was not an ancient concept," Dr. Miller said.

But only a few of the athletes, their oiled nude bodies glistening in the sun, would be taking victory laps.

In the Olympics of 448 B.C., Dr. Miller noted, two brothers from Rhodes each won a contest. They ran into the crowd to pick up their father, who had himself been an Olympic winner years earlier. As the three paraded triumphantly, spectators went wild and showered them with flowers. A Spartan shouted to the father: "Die now, Diagoras! You will never be happier."

By the fourth century A.D., with the spread of Christianity and the waning of belief in the ancient Greek gods, Dr. Miller wrote, the games "ceased completely to play any meaningful role in society."

It was not until 1896 that they were revived in their modern, international form, a tribute to the competitive spirit of ancient Greece.

After reading article 8, answer the following question: what was the significance of athletics for ancient Greeks?

Part 3: ROME

Augustus, First Emperor of Rome

© Filip Fuxa/Shutterstock.com

Roman Aqueduct (Pont du Gard, Southern France Near Nimes)

THE ROMAN EMPIRE

ANCIENT ROME

- By 2nd c. BCE, stronger power in Mediterranean Sea
- Why? Developed a more comprehensive concept of citizenship
- Offered partial or complete citizenship to the people they conquered (sometimes reluctantly but incrementally over several centuries)
- Way of co-opting elites of conquered lands

ENLARGEMENT OF CITIZENSHIP TO OTHER ETHNIC GROUPS

- As in Greece, citizenship was tied to military service
- However, it was not tied to ethnicity (Italy was conglomerate of different ethnic groups)
- Citizenship gave people legal protection and certain tax exemptions
  - Completing twenty-six years of military service in native units earned non-Roman soldiers citizenship
  - Citizenship could be passed to their descendants
  - In 3rd c. CE, emperor Caracalla granted citizenship to all free adult male of the empire

ROMANS BORROWED FROM CONQUERED PEOPLES

- Etruscans (gold and bronze metallurgy + toga + some gods + Greek alphabet in Etruscan form)
- Greek-style republicanism + deities + art, science, philosophy
- Persia (cult of Mithras)
- Egypt (cults of Osiris and Isis)

TWO MAJOR PERIODS OF ROMAN HISTORY

- The Republic (6th c. BCE-1st c. BCE) = Greek city-state (all adult males had rights and obligations in public life)
- The Principate (1st c. BCE-3rd c. CE) = individual rulers achieved autocratic command

THE ROMAN REPUBLIC

- Forum = political and civic center with temple and public buildings where leading citizens tended to government business
- Two consuls elected every year by Patricians (elite class by birth)
- Advised by Senate (also patricians) (+ ratified consuls' decisions)
- No debate as in Greece (voting only) + patron-client system

BROADENING OF POLITICAL PARTICIPATION (5TH-4TH C. BCE)

- Conflicts between Patricians + Plebeians (common people)
- Plebeians obtained right to elect their own officials (tribunes) who represented their interests
- Tribunes could intervene in all political powers + right to veto any measure they considered unfair + one of the consuls came from their ranks

FALL OF REPUBLIC

- Too much empire building
- Slaves and wealth poured into Rome
- To fight distant wars, citizen-soldiers had to leave their farms for many years
- Wealthy citizens bought lands of absentee soldiers + lands were consolidated into large estates (*latifundia*) tilled by slaves
- High concentration of slaves led to rebellions
- Ex-farmers fled to cities (no jobs)

FALL OF REPUBLIC

- Senate eliminated property qualifications for service
- Julius Caesar exploited this new policy and recruited virtually private armies offering pay, war booty, and free farms to landless men
- Professional fighters replaced citizen-soldiers
- Powerful generals competed for power

AUGUSTUS FIRST PRINCEPS

- Autocracy (imperial cult)
- Senate was preserved. It began issuing divine honors to deceased rulers
- People started revering the ruler as if he were a god (temples, altars, processions with his images)

PAX ROMANA (ROUGHLY 1ST TO MIDDLE 3RD C. CE)

- New cities emerged in Gaul, Germany, Britain, and Spain = Paris, Lyons, Cologne, London, Toledo
- More comfort (aqueducts + underground sewers + plumbing + public baths + swimming pools + gymnasia + circuses + stadiums + theaters)
- Invented concrete (provided increased strength and stability to buildings)
- Romans borrowed columns from Greek architecture but added arches, vaulted ceilings, enormous domed interior spaces
- Improved roads (two lanes, flat paving stones, mileage)

NEW LEGAL SYSTEM

- Jurists constructed an elaborate system of law that could apply to all people
- Established principle that defendants were innocent until proven guilty
- Ensured that defendants had right to challenge their accusers in court before a judge
- Judges could set aside laws that were unfair

ECONOMIC SPECIALIZATION

- Romans developed commercial agriculture and made economic specialization possible
- Instead of planting crops for local use, owners of *latifundia* concentrated on production for export
- Greece specialized in production of olive oil and grapevines
- Syria and Palestine = fruits, nuts, and wool fabrics
- Gaul = grain and copper
- Italy = pottery, glassware, bronze products

LINGUISTIC HOMOGENEITY

- Language of the Roman Empire (administration, justice) was Latin
- Many European languages borrowed a lot of vocab from Latin (French, Italian, Portuguese, Romanian, Spanish, even English)
- Greek remained language of science

FIRST WOMEN'S STRIKE

- Roman law gave male heads of families considerable authority
- But over time, women managed to exercise considerable power behind the scene
  - Earned right to own property, conduct monetary affairs, and even manage businesses
  - Lobbied for their husbands (senators)
  - Roman assemblies tried to restrict women's mobility (but women picketed the Senate and forced the repeal of the law)

CLICKER NOTES

THE ROMAN EMPIRE

1. Who were the Etruscans?

2. How did Greek travelers view the Etruscans and why?

3. What were the major achievements of the Roman Empire?

4. How did the Roman past impact Western civilization?

ROMAN ROADS

1. Description

2. Significance

ROME

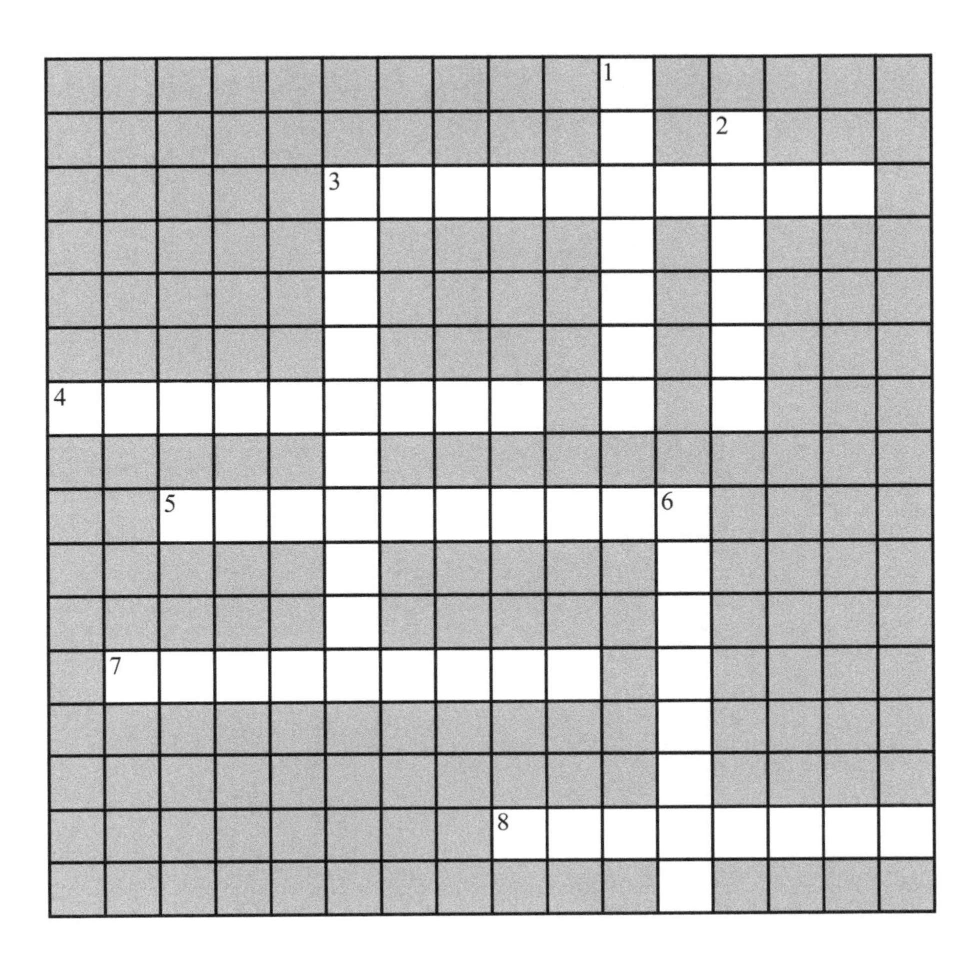

ACROSS

3. Constitutional monarchy of the early Roman Empire. It originally comes from a Latin word that means "first citizen"
4. Roman common people (plural)
5. Huge state-run and slave-worked farms in ancient Rome
7. Inhabitants of Northern Italy conquered by Romans. They helped convey Greek culture to their expanding conquerors
8. Form of government by elected representatives. In Latin, it means "public thing."

DOWN

1. Roman official of plebeian origin, elected by an assembly of plebeians to protect their interests
2. Famous Roman general who invaded the Gaul (modern France). Later roman emperors were called by the same name
3. Roman aristocrats and wealthy class (plural)
6. Semireligious title given to Octavian in 27 BCE and borne hereafter by all Roman emperors. It implied majesty and holiness

Chapter 12

The Ancient Mediterranean
Judaism and Christianity

The Dead Sea Scrolls

Part 1: FROM ANCIENT JUDAISM TO RABBINICAL JUDAISM

ANCIENT JUDAISM IN 6TH C. BCE (AXIS AGE)

- Exile in Babylonia (Axis Age)
- 586 BCE—Nebuchadnezzar II of Babylon
  - Destroyed Solomon's temple
  - Deported most of the population to Babylonia
- The people lost all that had defined them
  - Independence
  - King
  - Temple
  - Land

EFFECTS OF EXILE

- Diaspora
  - Babylon became a center of Jewish life
- Synagogue
  - Temple ritual was deemphasized
  - Written word took on new importance
  - Sabbath service of worship, study, psalms
- Scriptures—Oral traditions (= Oral Torah) were recorded
- Language
  - Hebrew declined, replaced by Aramaic

RETURN FROM EXILE

- Persian king Cyrus conquered Babylon
- Allowed Jews to return to homeland
- Jews rebuilt second temple

SECOND TEMPLE

- Priestly hegemony under Achaemenid Persians
- Priests developed second temple ritual
- Priests made final edition of Pentateuch (first five books of Moses = Written Torah)
- Priests compiled prophetic books (*Neviim*)

ALEXANDER THE GREAT IN 333 BCE

- Hellenistic culture
  - From Hellas, Greece
  - Major influence

- Alexandria, Egypt
  - Major Jewish center
  - Septuagint
- Judea
  - Hellenized upper class
  - Sought to transform Jerusalem into Greek city

PERSECUTION BY KING ANTIOCHUS IV (2ND C. BCE)

- Tried to destroy Judaism
- Forbade
  - Sabbath
  - Circumcision
  - Torah study
- Temple became shrine to Zeus

MACCABEAN REVOLT

- Antiochus's policies provoked revolt
- Priestly Hasmonean family led revolt
  - Judas Maccabee (the Hammer) = charismatic leader
- 164 BCE—Rebels had regained temple
  - Feast of Hanukkah
  - Recalls the rededication of second temple

ROMAN RULE (63 BCE-638 CE)

- 63 BCE—Roman general Pompey conquered Jerusalem
- Roman Empire rules Judea until Muslim conquest in 638 CE
- During Roman rule
  - Jews enjoyed some religious independence
  - Led by priests
  - Local government was entrusted to local princes, of whom Herod the Great (37 BCE-4 CE) was one
  - Herod was the ruler of Palestine at the time Jesus was born
- Later, procurators appointed by Roman emperors were placed in charge of smaller Palestinian territories

OUTSIDE INFLUENCES UP TO 1ST C. BCE

- Hellenistic culture
- Alien rule
  - Babylonian
  - Persian
  - Greek
  - Roman
- Diaspora
- New questions had arisen
  - What was authoritative scripture?
  - Who could interpret Scriptures?

FOUR IMPORTANT FACTIONS

- Sadducees
- Essenes
- Pharisees
- Zealots

THE DEVELOPMENT OF RABBINICAL/TALMUDIC JUDAISM

- Destruction of second temple in 70 CE
- Ended power of priesthood
  - Sacrifice was no longer possible
  - Ritual was thus deemphasized
- Judaism emphasized scripture interpretation
- Leadership passed from priests to rabbis

PHARISEES PROVIDED NEW LEADERSHIP. WHY?

- Politically reliable to Rome
- Pharisees established academy
  - Jamnia
- Pharisees = leaders of synagogue tradition

REINTERPRETATION OF JUDAISM AFTER FALL OF TEMPLE

- People of Israel is holy
- Every male head of Jewish household = priest
- Table in every Jewish house is an altar

FORMATION OF THE TALMUD

- 200 CE—Rabbi Judah compiled Mishnah (= repetition)
  - Codification of Torah in six sections
- Gemara (= tradition, completion)
  - Commentary of the Mishnah
  - Completed in 6th c.
- Mishnah + Gemara = Talmud

TWO TALMUDS

- Palestinian (or Jerusalem) Talmud (about 400 CE)
- Babylonian Talmud (about 600 CE) = most authoritative
- Contain
  - legal material, commandments, rules for living
  - anecdotes, stories, tales

CLICKER NOTES

The Beginnings

1. What is the problem of defining Judaism?
2. Is Judaism pluralistic?
3. When did Judaism begin?
4. Where did Abraham live? What happened to him?
5. What was the greatest contribution of Judaism?
6. What is a covenant?
7. What was the Abrahamic covenant?
8. What happened to Israel in Egypt? How did they escape from Egypt?
9. What did Moses receive from God at Sinai?
10. What does the word Torah mean?
11. What did the twelve tribes do?
12. Who was the first king? Who succeeded him?
13. Who built the temple? When?

14. What was the center of the cult?

15. What was the promised land?

16. What were the people of the united kingdom called?

17. After Solomon, what happened to Israel?

18. How was the Northern Kingdom destroyed?

19. When did the Babylonians invade Judah?

20. What did Cyrus do for the Jews?

21. When did Jews build the second temple?

22. The repatriates canonized which part of the Bible?

23. What are the three parts of the Hebrew scriptures?

24. What did the Maccabees do? When did they live?

25. When did Romans conquer Jerusalem?

26. What was Judaism's greatest crisis? What happened to Jerusalem and the temple? When?

After the Destruction of the Second Temple

1. Who destroyed the second temple? When was it destroyed?

2. What was the consequence of the destruction of the second temple?

3. Where did Jews live after the destruction of the temple?

4. Instead of being temple centered, Judaism became __________ centered. Which period of Jewish history does this remind you of?

5. When did Jews visit the temple previously?

6. What was the synagogue?

7. How was Yahweh worshiped before the destruction of the temple?

8. What is the Mishnah? When was it compiled?

9. What is the Talmud? When was it compiled?

10. How and where did Jews spread in the world?

11. When did Christianity arise? What was its consequence?

12. How did Jews cope with exile?

RABBINICAL JUDAISM

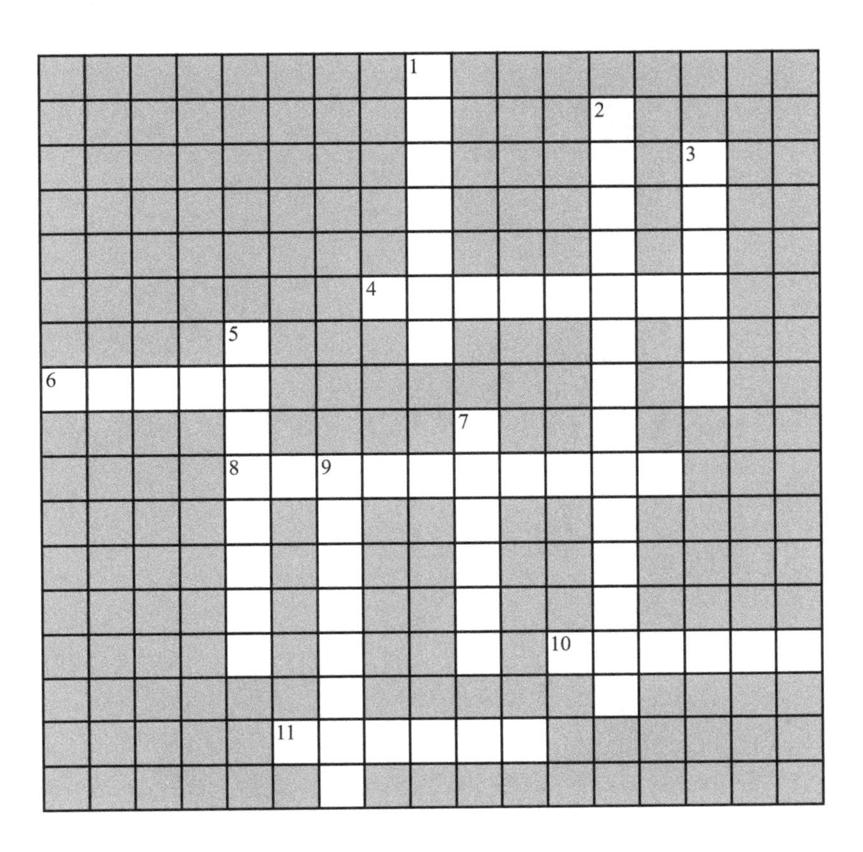

ACROSS

4. Priestly faction influential during the Second Temple period
6. Teacher
8. Greek translation of the Hebrew Bible
10. Codified oral Torah (Mishnah + Gemara)
11. Anti-Roman, nationalistic Jewish faction

DOWN

1. Literally, "Repeated Study" or oldest authoritative postbiblical collection and codification of the oral Torah
2. Babylonian king who destroyed the first temple in Jerusalem (6th c. BCE)
3. A reclusive semi-monastic Jewish group on the Dead Sea
5. Dispersed people (in Babylonia, Persia, Egypt, etc.)
7. Town near the Mediterranean coast where Romans allowed Pharisees to open an academy of rabbinical study after the fall of the second temple
9. Faction that emphasized strict observance of the written Torah and oral Torah

Part 2: Christianity

© Antony McAulay/Shutterstock.com

Jesus on One of the Mosaics that Adorn the Former Byzantine Basilica Hagia Sophia (Istanbul, Turkey)

CONSTANTINE THE GREAT

© Oleg Senkov/Shutterstock.com

Significance: Why is Constantine important to remember?

____

____

____

____

CHRISTIANITY

BASIC INFO ABOUT CHRISTIANITY

- Faith in Jesus of Nazareth
- Jesus is the Christ (Messiah = Anointed)
- God became human in Christ to save humanity
- Jesus provided a new interpretation of the Torah
- "Sabbath was made for man"
  - Jesus' disciples picked grain
  - Jesus healed
- Jesus rose from the dead

UNIQUE CHRISTIAN CONCEPT OF ORIGINAL SIN

- Sin = failure to live in harmony with God
- Original sin
  - Human will has become corrupted
  - Adam's disobedience brought sin + death
- Jesus = Savior (Messiah)
  - Recreated a good will
  - Repaired the world through his sacrifice

WHY DID JEWS BRING SACRIFICES TO THE TEMPLE?

- To thank God
- To praise Him
- To cleanse their sins (Atonement)
- Christians believe that Jesus was God's sacrifice + fulfillment of Hebrew prophecy

CHRISTIANITY: BEGINNINGS

- Jesus = fulfillment of Hebrew prophecy
- In Matthew's Gospel
  - Jesus = son of David, son of Abraham = Messiah
  - Virgin birth
  - Matthew quotes Isaiah:

    "Behold, a virgin shall conceive and bear a son, and his name shall be called Emmanuel (which means, 'God with us')"
  - Jesus' family escapes from the slaughter of infants by King Herod = Exodus account of Israelites' escape from Egypt
  - Jesus = "New Moses" = Sermon on Mount = Torah on Mount Sinai

RABBINICAL JUDAISM AND CHRISTIANITY

- Both Christianity and Modern Judaism are two developments of ancient Judaism
- Both Christianity and Modern Judaism developed by reflecting on the destruction of the second temple in 70CE

- Judaism developed the Torah (teaching)
- Christianity developed Christology

THE JESUS MOVEMENT = A FORM OF JUDAISM

- Christianity began as a Jewish movement
- Jesus was a Jew
- Jesus taught as a Rabbi [master], a teacher of Jewish law
- Jesus' followers were initially all Jews
  - Worshiped with other Jews in the temple
  - Observed the Torah and Jewish feasts
- Jesus' followers came to believe
  - Jesus was promised Messiah
  - Jesus had risen from the dead

EARLY CHRISTIAN RITUALS

- Based on Jewish rituals
- Baptism = *mikveh* [ritual bath]
  - Christian rite of initiation
- Eucharist = thanksgiving
  - Reenactment of Jesus' last meal with disciples
  - Similar to Passover meal

HOW DID CHRISTIANITY SEPARATE FROM JUDAISM (OUTLINE)

- BEFORE THE DESTRUCTION OF SECOND TEMPLE (before 70 CE)
  - "God-fearing" Gentiles (= non Jews) enter Christian community (30s CE)
  - Council of Jerusalem (48 CE) = circumcision unnecessary for Christians
- AFTER DESTRUCTION OF SECOND TEMPLE (70 CE)
  - The leadership becomes increasingly gentile

BEFORE THE DESTRUCTION OF SECOND TEMPLE = PAUL OF TARSUS, APOSTLE TO GENTILES

- Paul of Tarsus = learned Jew
- Roman citizen
- He initially persecuted Christians
- Dramatic conversion after vision of Jesus
- Became itinerant Christian missionary
- Letters to churches became part of New Testament

COUNCIL OF JERUSALEM (48 CE)

- Critical issue = status of Gentile converts
- GENTILES = NON JEWS
- Conservative Christians argued that Gentiles must
  - be circumcised
  - follow Mosaic law

- Paul argued
  - faith in Jesus Christ—not obedience to Torah—saves
  - Gentiles should be exempt from circumcision
  - But still keep moral commandments of Mosaic covenant
- Conclusion—Gentile Christians did not need to be circumcised

AFTER THE DESTRUCTION OF SECOND TEMPLE = MAIN CONSEQUENCES

- Jewish sacrificial system is at an end
- How is it possible to reach God?
- Christians develop Christology
  - Four gospels are written after 70 CE
  - Jesus = sacrifice
  - Destruction = God's punishment of non-Christian Jews
- Jews develop Torah to keep people of Israel holy. Through Israel other nations will come to know God.

CHURCH BECAME INCREASINGLY GENTILE

- New Testament is in Greek, not Aramaic

THE NEW TESTAMENT

- Testament = covenant (contract)
- Old Testament = Hebrew Bible (Pharisaic *TANAKH)* = *SEPTUAGINT* (Greek translation of the Hebrew Bible)
- The New Testament has twenty-seven books. Among them
  - four gospels (canonical/authoritative Gospels)
    [Synoptic Gospels = contain similar stories = Mark, Matthew, and Luke]
  - Apostle Paul's letters (earliest documents)

ORGANIZATION OF CHURCH

- Christians initially expected imminent return of Christ
- When Christ tarried, Christians organized Church (*ekklesia* = gathering, convocation)

EPISCOPAL ORGANIZATION

- Bishop (*episkopos*, overseer)
  - Leader of urban Christian community
- Bishop of Rome = principal bishop
  - Apostle Peter = first bishop of Rome
  - By 4th c., bishop of Rome is called Pope (father)

ROMAN PERSECUTION

- After Rome burnt in 64 CE, Nero accused Christians of starting the fire
- Labeled them enemies of the state

INSTITUTIONALIZATION OF THE CHURCH

- Emperor Constantine (r. 306-337)
- Edict of Milan or Edict of Toleration (313)
  - Legalized Christianity
- Made Sunday a legal holiday
- Moved capital to Byzantium

MORE ON CONSTANTINE

- Sought to unify empire through religion
- Called the first *ecumenical* council to resolve Church conflicts
  - Ecumenical = from the whole world
- Later emperors called similar councils

CONCEPT OF TRINITY = ONE GOD

- What did Jesus mean when He said:
  - "No one can come to the father except through me" (John 14:6)
  - Or "I and the father are one" (John 10:30)
- What is the relationship between
  - God, the Father
  - Jesus, the Son
  - and the Holy Spirit?
- How can polytheism be avoided?

WHAT DOES GOD DO IN JUDAISM, CHRISTIANITY, AND ISLAM?

- Creates
- Redeems
- Reveals

IN CHRISTIANITY

- God is the Father (Creator), the Son Jesus (the Redeemer), and the Holy Spirit (the Revelator)
- The Trinity refers to God's activity in relation to humanity
- It explains how God works through Christ
- Metaphor used by early Christians to explain Trinity: The Sun is light and heat

IMPORTANT EARLY ECUMENICAL COUNCILS

- First ecumenical council in Nicea, 325 CE
  - Declared that Christ is fully God
  - Arius lost: he believed that Jesus was not eternal but created within time
    - Arius was exiled
    - He continued proselytizing
    - Converted Germanic tribes to Christianity
  - Bishop Athanasius won
  - The Father and the Son were coeternal
  - Only God could repair the world

- Second ecumenical council in Constantinople, 381 CE
  - Formulated Doctrine of Trinity
  - God has three personas
  - Persons are NOT beings
  - Persons = God's activities or manifestations
  - Trinity = Monotheism
  - Jews and Muslims misunderstand concept of Trinity as form of polytheism
- Fourth ecumenical council in Chalcedon, 451
  - Declared that Christ is fully human and fully God

JEWS AND CHRISTIANS

- In 381 (under Emperor Theodosius) Christianity = state religion
- All pagan worship illegal
- Judaism permitted but under severe legal restrictions
  - Jews could not own slaves
  - Jews could not marry non Jews
  - Jews could not proselytize anymore
  - Jews could not build new synagogues
  - Jews could not work for the government
  - Jews could not teach in public institutions
  - Jews could not serve in the army

CLICKER NOTES

CONSTANTINE

1. Who was Constantine?

2. What kind of vision did he have?

3. When did it occur?

4. What is the evidence that Constantine genuinely converted to Christianity?

5. What is the evidence against his conversion?

6. How did the conversion of Constantine affect Christianity and the Roman Empire?

CHRISTIANITY

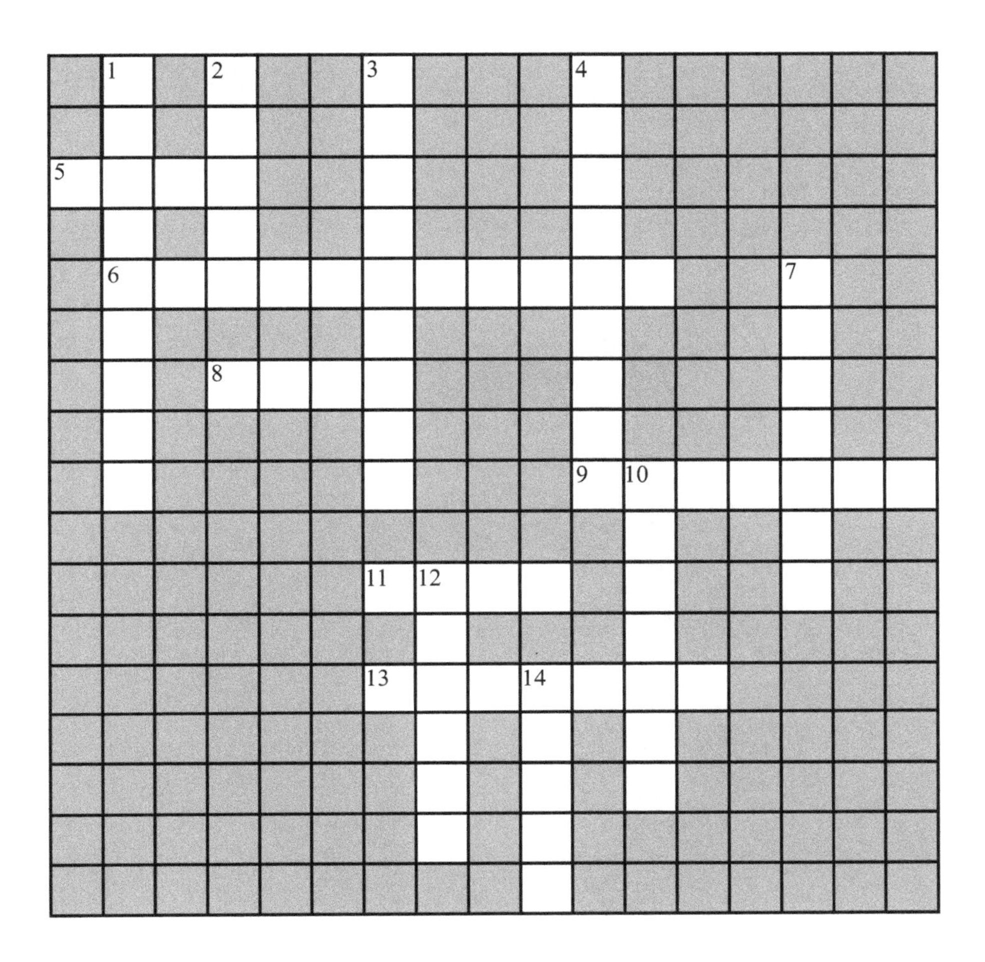

ACROSS

5. A Jew from the Greek city of Tarsus in Anatolia who called himself the "Apostle to the Gentiles"
6. First Roman emperor to legalize Christianity
8. Bishop of Rome and vicar (agent) of Christ
9. Anointed
11. Image or stylized wooden painting depicting Christian holy persons
13. God in three persons

DOWN

1. In this city (now called Kadiköy in Turkey), it was declared that Christ is fully human and fully God
2. City in Italy where Christianity was legalized
3. Covenant in Greek
4. Constantine renamed this small village Constantinople (now called Istanbul in Turkey)
7. Non-Jew
10. Letter
12. Messiah in Greek
14. In this city (now called Iznik in Turkey), it was declared that Jesus is fully God

Chapter 13

Islam

The Kaaba in Mecca (Saudi Arabia)

FORMER MOSQUE OF CORDOBA IN ANDALUSIA, SPAIN

Significance: Why is it important to remember the contributions of Islamic Spain?

FAMOUS CENTRAL ASIAN PHYSICIAN AND PHILOSOPHER: IBN SINA (AVICENNA) 10TH-11TH C.

Portrait of Avicenna on a Postage Stamp (USSR, 1980)

Avicenna's Tomb in Hamedan, Iran

Significance: Why is it important to remember Avicenna?

____

____

____

____

WHERE ARE MECCA AND MEDINA ON THIS MAP?

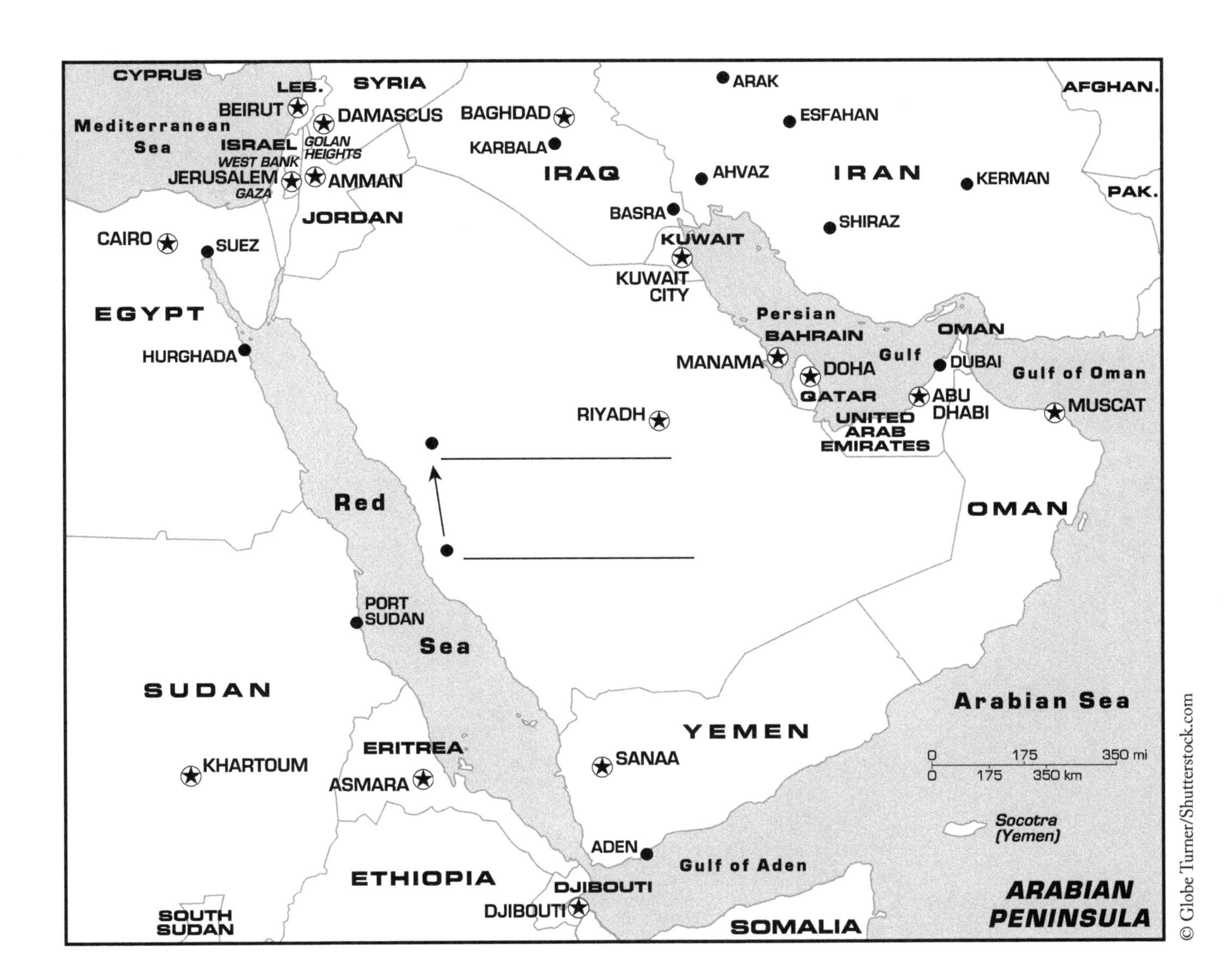

How is the journey from Mecca to Medina called in Arabic? ______________________________

ISLAMIC CIVILIZATION

WHAT DO THESE WORDS MEAN?

- Islam = submission to God
- The followers of Islam are called Muslims (= ones who submit to Allah)
- Allah = God in Arabic

JUDAISM, CHRISTIANITY, AND ISLAM

All three religions come from the Middle East

- ***SIMILAR BELIEFS***
  - One god
  - Patriarch Abraham
  - Universal prophetic message
  - Angels
  - Satan
  - Divine judgement
- ***DIFFERENCES***
  - Muslims believe that Jews and Christians have gone astray
  - God sent Muhammad to restore the true faith of Abraham

ARABIA AT THE TIME OF THE PROPHET'S BIRTH

- Urbanized, trading community in the Saudi desert
- Polytheistic
- Allah, high God of the Kaaba
- Exposed to Jewish, Christian, and Zoroastrian influences

MUHAMMAD

- Birth in Mecca
- 7th c. = first revelation
- *Hijra* (flight to Medina)
- Capitulation of Mecca
- Death of Muhammad

REVELATION OF QUR'AN

- 610 CE = Muhammad was forty years old
- Received first revelation in a cave at Mount Hira
- Jibreel (Gabriel) appeared and ordered Muhammad to recite
- This time is celebrated during "Night of Power and Excellence" [Ramadan]

HIJRA

- Opposition in Mecca forced Muhammad and followers to retreat to Medina

- In Medina, Muhammad combined the role of
  - religious prophet
  - political ruler

DEVELOPMENT OF QUR'AN

- Received divine revelations over twenty-two years (610-632 CE)
- Qur'an = recitation
- Ca. 650 CE, third caliph (Uthman) established authoritative text
- Uthman's text is the one still used

COMPOSITION OF QUR'AN

- 114 chapters (*SURAS*) of 6,000 verses
- Chapters arranged according to length, not chronology

THE QUR'AN DURING THE MECCAN PERIOD

- For Muslims, the Qur'an is the Book of God
- Earliest suras are the shortest
- Earliest suras at end

THEMES OF EARLY MESSAGE IN MECCA

- God is just, all-powerful, majestic, and holy
- God's judgment of humanity at end of the world
- Resurrection of dead to
  - eternal felicity
  - eternal punishment
- God
  - demands worship, praise, gratitude
  - demands moral behavior
  - condemns female infanticide
  - emphasizes generosity toward others, particularly the poor, the outcast
  - is merciful and compassionate

THEMES IN MEDINA

- *Umma* = community of a prophet
- In Medina, *Umma* = community of Muhammad
- Revelations become longer
- Revelations include legislation, community affairs

VIEWS OF THE QUR'AN

- Uncreated Word of God
- Untranslatable
- Only undistorted revelation
- Restoration, not reformation
  - Adam was the first prophet

  - Every child is Muslim
  - No "original sin"

JUDAISM VERSUS ISLAM

- Until 624, Muslims prayed toward Jerusalem
- Muslims shared fasts and prayed same number of times
- But Jews denied Muhammad's prophethood
- Muslims started praying toward Mecca
- Kaaba understood as originally built by Abraham (Ibrahim) and Ishmael (Ismail)

THE RIGHTLY GUIDED CALIPHS (632-661 CE)

- 632—Muhammad died
- No clear successor to Muhammad
- Muslim elders created institution of caliphate
- Caliph (*Khalifah*) = representative of another (Muhammad)
- Caliphate modeled on tribal shaykh or chief
  - First among equals
  - Moral authority

QUESTIONS OF SUCCESSION

- Should caliph be member of Prophet's tribe, the Quraysh? [Umayyad clan]
- Should caliph be chosen from Muhammad's immediate family? (Shiites say YES, Sunnis say not necessarily)
- For Shiites, the *first* successor to Prophet Muhammad should have been Ali (cousin + son-in-law)

RIGHTLY GUIDED CALIPHS (632-61 CE)

- Abu Bakr
- Umar
- Uthman (belonged to Umayyad clan of Quraysh tribe)
- Ali (Prophet's cousin + son-in-law)

ACHIEVEMENTS

- Abu Bakr: conquest of Arabia
- Umar and Uthman: conquests beyond Arabia (parts of Byzantine Empire + Persia)
- Definitive edition of Qur'an under Uthman

ALI (FOURTH RIGHTLY GUIDED CALIPH)

- Muhammad's cousin
- He had also married Fatima, Muhammad's daughter
- So his sons, Hasan and Husayn, were Muhammad's grandsons

CIVIL WAR AND SCHISM OF SUNNI AND SHI'I

- Civil war broke out with assassination of Uthman
- Civil war between Ali and Muawiya, the leader of the Umayyad clan of Quraysh tribe

- Former supporters assassinated Ali because Ali favored negotiation
- Shiites held that succession belonged to Ali's two sons, Hasan and Husayn, but Umayyads rejected this idea

TRAGIC EVENT AT KARBALA (IRAQ) 680 CE

- Main consequence: Umayyads held power until m. 8th c. (time of Abbasid Revolution)
- This led to most significant division in Islam: Sunni/Shiites
- Day of Ashura (= the Tenth)
  - Commemorates martyrdom of Husayn, Ali's son, the Prophet Muhammad's grandson

SHIISM

- Shi'i = party of the House of Ali
- Sunni = those who follow the Prophet's example [the Sunna of the Prophet]
- Shiites emphasized the authority of genetic connection to prophet
- Shiite views of leadership
  - Leader should be descendant of Ali and Husayn
  - Leadership vested in Imam (leader)
  - Imam is not a prophet
  - Imam is
    - divinely inspired
    - sinless
    - infallible
    - a religio-political leader
    - a direct descendant of the Prophet
    - the final authoritative interpreter of God's will

IMAMS

- In Sunni Islam = prayer leader
- In Shiite Islam
  - Descendant of Muhammad
  - Divinely inspired
- Imams are NOT Ayatollahs
- Ayatollahs = scholars of Islam = interpret sacred texts in the name of "Hidden Imam"

WHERE CAN SHIITES BE FOUND?

- Iran
- Iraq
- Syria
- Lebanon
- Saudi Arabia
- Yemen
- Pakistan

MAJOR ACCOMPLISHMENTS UNDER UMAYYADS (661-750 CE)

- Conquered North Africa + southern Spain
- Capital city: Damascus
- Kept local administration in place
- No forced conversion (soldiers confined to garrisons)
- Conversions did happen but new converts not treated as equals, leading to downfall

ABBASIDS (750-1258 CE)

- Rebellion led by Persian converts to Islam
- Capital city: Baghdad
- Development of Islamic law + Islamic philosophy
- Sufism

THE ABBASID EMPIRE

- Islam spread in different ways
- Conquest? Escape from tax on non-Muslims? Not so sure (no forced conversion, early converts still had to pay the tax)
- Trade (silk roads in Eurasia + trans-Sahara roads in Africa + sea routes of the Mediterranean Sea and Indian Ocean)
- Slavery + marriage
- Teaching (Sufism)

CONVERSION AND SOCIOECONOMIC IMPACT

- Conversion-related migration triggered urban growth
- Conversion meant that people had to leave their original communities
- New converts left for Arab military settlements
- Military garrison Fustat became Cairo

SHARIA (ISLAMIC LAW)

- Guide, path
- Sources
  - Qur'an
  - *Sunna* of Prophet (contained in hadith)
  - Analogy
  - Consensus
- Legal diversity = Four *Madhabs* (legal schools) in Sunni Islam

SUNNA

- In pre-Islamic times referred to customs and usages of elders
- At the death of the Prophet, referred to the "life-example" of Muhammad's word and deed
- Principal source: Hadith literature (*hadith* = "recollection," remembrance of act and saying of Muhammad)
- Guide of proper conduct for all Muslims

FIVE PILLARS OF ISLAM

- Statement of Faith [*shahada*]
- Prayer [*salat, namaz*]
- Alms [*zakat*]
- Fast [*sawm*]
- Pilgrimage [*hajj*]

ISLAMIC MYSTICISM (SUFISM)

- Sufi = rough wool = garment worn by early Sufis as part of ascetic practice
- Goal of Sufism = unity with God
- Sufi practices to achieve this goal
  - *Dhikr* (zikr) = communal recitation of names of God
  - *TARIQA* = Sufi order = important for spread of Islam
- Important role in spread of Islam
  - Music and poetry
  - In vernacular (local) languages

GLOBAL IMPACT OF ISLAMIC CIVILIZATION

1. TRADE

- Revived silk road networks from China to the Mediterranean Sea
- More sophisticated camel saddles
- New caravanserais
- Expanded innovations in nautical technology
  - Compass from the Chinese
  - Lateen sail from Asian and Indian mariners
  - Astrolabe from Greeks and Romans

2. PAPER REVOLUTION

- Chinese craftsmen made paper since 1st c. CE
- But technology did not spread beyond China until Muslim conquests
- Books could be made in larger quantities

3. AGRICULTURE

- Transfer of new crops, fruits, vegetables from East to West
- This led to richer and more varied diet
- New crops had industrial use (cotton)
- Publication of numerous manuals about methods of irrigation, fertilization, crop rotation
- Increase of agrarian surpluses

4. ECONOMY

- Banks (lent money + served as brokers for investments + exchanged different currencies)
- Much larger scale (different branches throughout the Islamic world)
- Honored letters of credit called *saqq* (= check)

5. MEDICINE

- First hospitals emerged in Muslim Persia, Syria, and Egypt (separate wards for particular cases, dispensary, and library)
- Muslim physicians
  - Diagnosed cancer of stomach
  - Prescribed antidotes for poisoning
  - Discovered that bubonic plague could be transmitted by clothes
  - Invented dental instruments
  - Performed cataract surgery, caesarians
  - Treated mental illnesses

6. ISLAM AND GREEK PHILOSOPHY

- Muslim revived and preserved ancient Greek works
- Ibn Rushd (Averroes)— 12th c. (Cordoba, Spain)
  - Wrote learned commentaries of Aristotle
  - Prized by Christian theologians

AVICENNA [IBN SINA] (11TH C., CENTRAL ASIA)

- Famous physician and philosopher
  - No contradiction between science and religion
  - No contradiction between religion and philosophy
- Discovered contagious nature of tuberculosis
- Diseases can spread through water and soil
- His books were studied in University of Paris up to 17th c.

WOMEN

- Seclusion + veiling already existed in Byzantine (Christian) and Persian (Zoroastrian) Sassanid/ Sasanian Empire
- Only slaves had no hat, no veil
- Women had rights
  - Could divorce under specific conditions
  - Could own property and inherit
  - Could practice birth control
  - Could testify in courts or go on pilgrimage

SLAVERY

- Muslims engaged in slave trade
- Muslim could not enslave other Muslims or People of the book
- Exception = prisoners of war
- No hereditary slave society
- Slaves converted to Islam and masters freed them as an act of piety
- Offspring of slave women and Muslim men were born free

CLICKER NOTES

THE PROPHET MUHAMMAD'S LIFE AND BACKGROUND

1. How many times does the call for prayer sound? What does the muezzin (the caller for prayer) say?

2. How many people on earth are Muslim?

3. What were some of Islam's contributions to Western civilization?

4. Who was the founder of Islam?

5. Gather as much biographical data as you can about the Prophet from the video. (When was he born? Where? What was his profession? Who was the Prophet's first wife? Why was he called the Trusted One? What happened in a cave near Mecca?)

6. What can you say about pre-Islamic Arab society and religions?

7. What was the spiritual and economic importance of the Kaaba?

8. What was the Prophet's message? What were the social implications of his message?

9. Was Muhammad a poet? What does the Qur'an say?

10. What is the Qur'an? How was the Qur'an put together? What was it about? How is God represented in the Qur'an?

11. How is Muhammad represented in Islam? Are the miniatures you saw devotional images?

12. Was Muhammad immediately accepted? If not, who challenged him and why?

13. Why was Muhammad invited to Yathrib? What is the current name of Yathrib and what does this mean?

14. What does the word Hijra mean? When did it occur? And why?

15. What distinguished the Prophet's first community?

16. Did Islam challenge other religions?

17. How do Muslims pray and what does this form of worship symbolize?

18. When was Mecca conquered? What did Muhammad do when he entered Mecca?

ISLAM: EMPIRE OF FAITH

1. In pre-Islamic Arabian society, who was responsible for linking the various tribes to their ancestors and celebrating their ancient values?
   A. Animist priests
   B. The poets
   C. The warrior chiefs
   D. The scribes

2. Pre-Islamic Arabian society was often torn apart by:
   A. Byzantine conquests of Arabian territory
   B. Sassanian conquests of Arabian territory
   C. Tribal blood feuds
   D. The sudden influx of trade caravans

3. According to the film, what was the simple yet radical proclamation of the Qur'an?
   A. Muhammad was the incarnation of God
   B. Muhammad was the Son of God
   C. There is only one God
   D. Jesus was a false prophet

4. Since images and representations of God or the prophets are generally forbidden or frowned upon, what is the traditional form of artistic expression in Islam?
   A. Music
   B. Dance
   C. Arabic calligraphy
   D. Sculpture

5. When Muhammad conquered Mecca, what did he do to the animist ("pagan") population of the city?
   A. Executed the men, and took the women and children as slaves
   B. Granted the people amnesty
   C. Expelled the people from Mecca, forcing them to live off in the desert as nomads
   D. Compromised his earlier teachings, and accepted some of the animists' beliefs

THE UMAYYAD AND ABBASID EMPIRES

1. Arab expansion:

   a. Geography

      Name cities on the map

   b. Timing

2. Why did Arabs expand successfully?

3. How did Arabs rule?

4. How did Arabs transform the conquered lands?

5. What was Islam's great work of art? Where is it located?

6. How did ideas/inventions spread in the new Arab Empire?

7. Baghdad (currently capital of Iraq). In 750, the Abbasid dynasty moved the capital of the empire from Damascus to Baghdad. Describe the city.

8. What were the needs of the new Empire?

9. What were the Islamic original contributions to the world? And in particular, the West?

10. Describe Cordoba (Spain) under Arab rule (711-1236). Why did Cordoba impress Northern Europe travelers?

11. When did the first Crusade occur? And why?

12. What was the name of the pope who launched the Crusades?

13. What impact did the Crusades have on the Western mind and culture?

14. Did Christendom triumph?

15. Compare the taking of Jerusalem by the Crusaders and Salah al-Din's in 1187.

16. How do you think Muslims look at the Crusades?

THE FIRST FOUR CALIPHS

1. Abu Bakr	a. Jurists and scholars of Islam
2. Ali	b. Third caliph
3. Fatima	c. Annual Shii festival to commemorate the martyrdom of Husayn
4. Husayn	d. Grandson of Muhammad and second son of Ali
5. Uthman	e. Muhammad's daughter
6. Ashura	f. First caliph or successor to Muhammad (also called "Rightly Guided")
7. Ulama	g. Second caliph
8. Umar	h. Fourth caliph, Muhammad's cousin and son-in-law

Please format your answers as (1,a) below:

HOW FLUENT IS YOUR ARABIC?

1. Shahada	•	•	a. Chapter of the Qur'an
2. Umma	•	•	b. Legal school
3. Ulama	•	•	c. Example of the Prophet
4. Sunna	•	•	d. Witness of faith
5. Hadith	•	•	e. Record of the deeds and sayings of the Prophet
6. Sharia	•	•	f. Scholars of Islam
7. Madhab	•	•	g. Institution of higher learning
8. Hijra	•	•	h. Community of the Prophet
9. Sura	•	•	i. Journey from Mecca to Medina (Yathrib)
10. Madrasa	•	•	j. Islamic law

Please format your answers as (1, a) below

TIME OF THE PROPHET MUHAMMAD

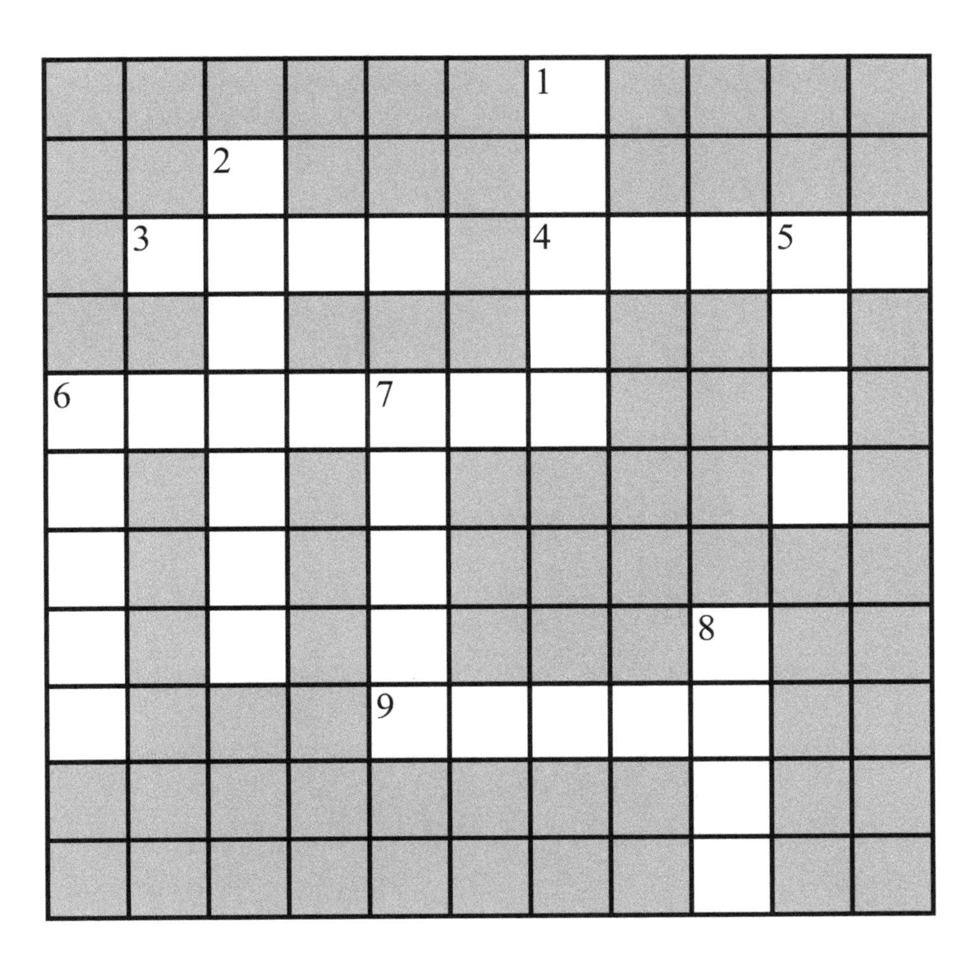

ACROSS

3. Chapter in Qur'an
4. According to Islamic tradition, this prophet did not die on the cross
6. The Prophet Muhammad's first wife (business-woman and mentor)
9. City where the Prophet Muhammad was born

DOWN

1. Journey from Mecca to Yathrib (later called Medina)
2. Tribe of the Prophet Muhammad
5. Community of the Prophet
6. The "cube" in Arabic, shrine originally built by Abraham (Ibrahim) and Ishmael (Ismail)
7. Submission in Arabic
8. Pilgrimage to Mecca

THE CALIPHATE

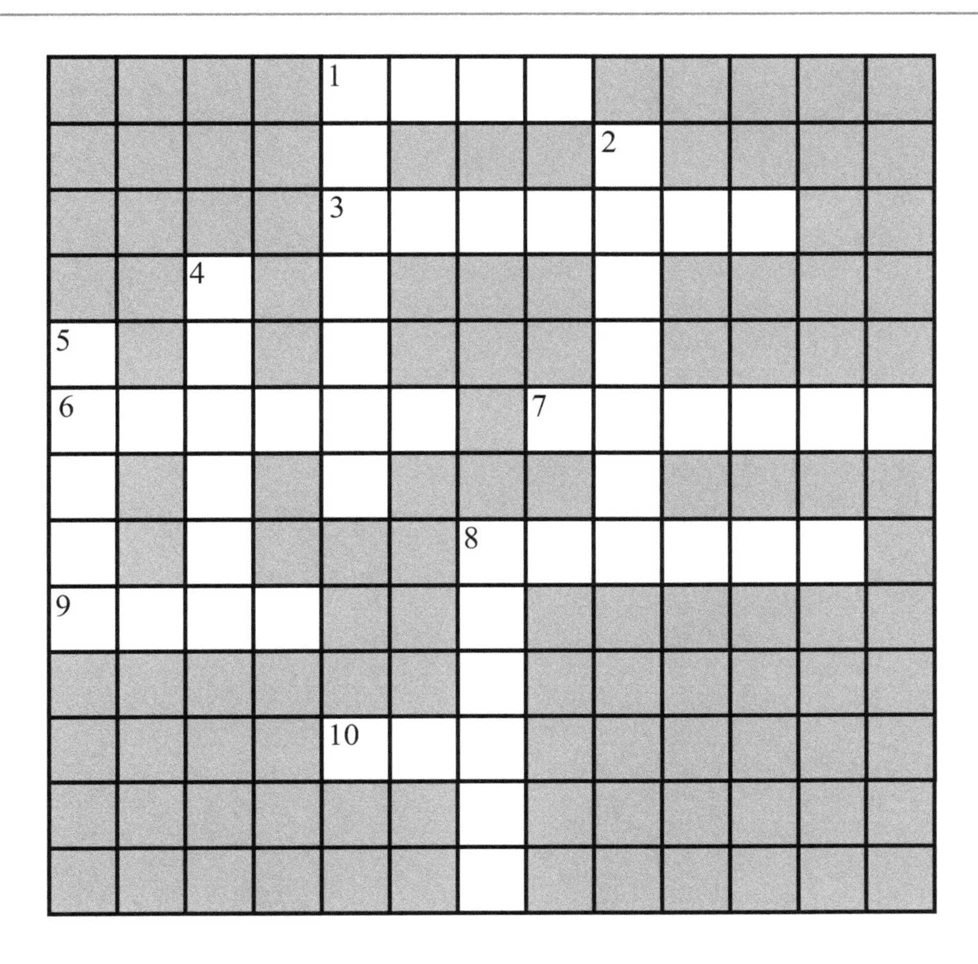

ACROSS

1. This caliph took Jerusalem
3. This caliph was responsible for consolidating the rule of Islam in Arabia
6. This caliph is responsible for ordering the compilation of the Qur'an as it is today
7. Deputy or successor of the Prophet Muhammad as religious and political leader of the Muslim community
8. Arabic word for a tribal elder
9. Prayer leader in Sunni Islam and direct descendant of the Prophet in Shiite Islam
10. For Shiites, the Prophet Muhammad chose this caliph as his successor

DOWN

1. Famous clan of the Quraysh tribe in conflict with the Party of the House of Ali (Shiites) and founder of the very first Muslim empire
2. Famous battle in Iraq where the Prophet's grandson, Husayn, was killed
4. This day commemorates the death of Husayn, son of Ali and grandson of Muhammad, at the battle of Karbala.
5. Majority of Muslims claiming to follow the example of the Prophet and holding that succession to the Prophet did not depend on hereditary descent from Ali
8. Minority of Muslims, following the Party of Ali

THE EXTERIOR AND INTERIOR PATHS OF ISLAM

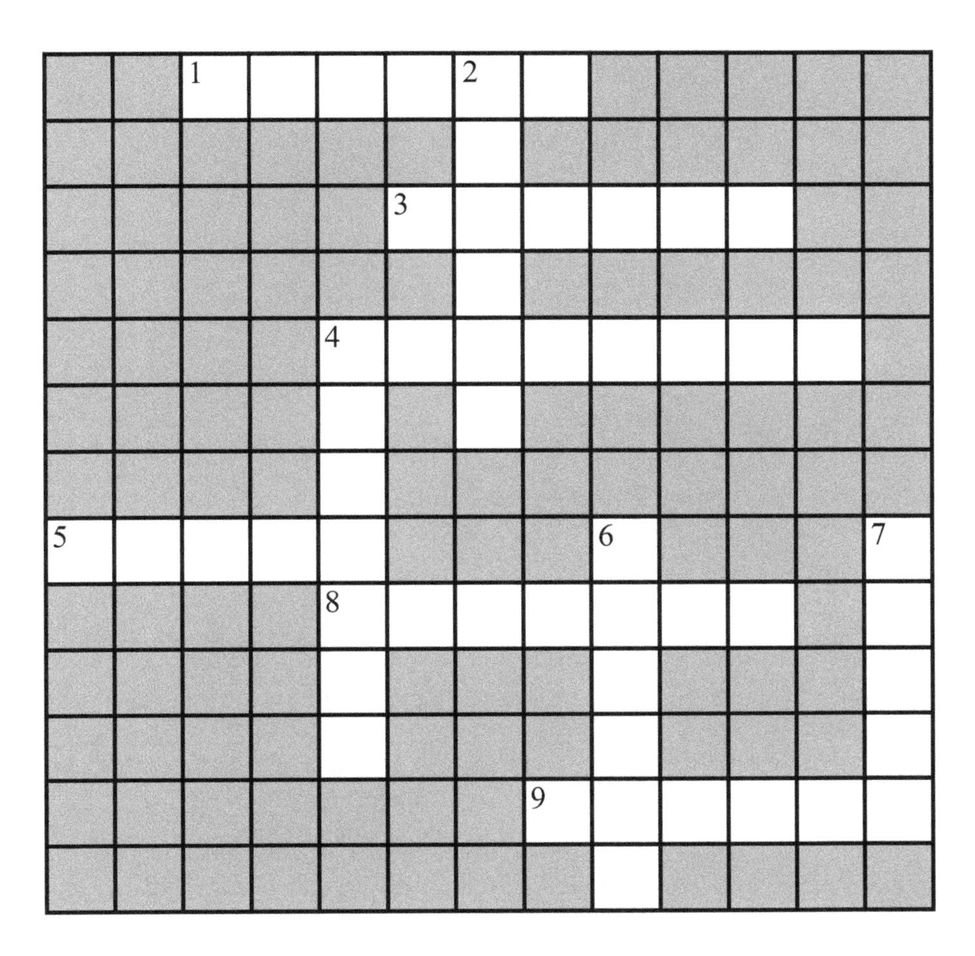

ACROSS

1. Islamic mysticism that advocates direct union with God through prayers, contemplation, and/or music.
3. Record of the deeds and sayings of the Prophet Muhammad
4. Famous Central Asian philosopher and physician
5. Religious scholars
8. Witness of faith
9. Sufi order

DOWN

2. The "path" in Arabic or Islamic Law
4. Second great Islamic empire that overthrew the Umayyad caliphate
6. Legal school
7. Example of the Prophet Muhammad

Chapter 14
The Two Worlds of Christendom

Notre Dame, Paris (Gothic Cathedral, 12th c.)

St George, an Icon of the 12th Century (Moscow, Russia)

Icons (Mary and Jesus)

MAJOR SPLIT BETWEEN EASTERN AND WESTERN CHURCHES

FIRST MAJOR SPLIT IN CHRISTIANITY (11TH C.)

- Split between Orthodox (Eastern) and Catholic (Western) churches
- Mutual excommunication

WHAT CAUSED THE DIVORCE?

- Question of leadership within church
- Are all bishops equal?

DIVISION OF ROMAN EMPIRE

- 330 (4th c.) Constantine moved capital to Constantinople
- Two halves grew apart

MAIN DIFFERENCES BETWEEN EAST AND WEST

- Language
- Church-state relations
- Episcopal authority
- Liturgical practices

LANGUAGE

- *WEST* = Latin
  - Rome had *one* sacred language, Latin, until 1960s
- *EAST* = linguistic diversity
  - Primary Christian language in East was Greek
  - Orthodox Church always used several sacred languages
    - Greek Orthodox use Greek
    - Slavic Orthodox use Slavonic
    - Arabic Orthodox use Arabic
    - Iraq and Palestine Christians use Syriac

CHURCH-STATE RELATIONS

- *BYZANTIUM*
  - Byzantine Empire lasted much longer in East
  - 1453 (15th c.) last emperor in Constantinople
  - Emperor ruled over both church and state in the name of Christ
- *WESTERN EUROPE*
  - Roman Empire in West lasted shorter time
  - 476 (5th c.) last emperor in Rome
  - Pope took on greater authority
  - Church developed as an independent, competing organization vis-a-vis states

PAPAL PRIMACY

- Orthodox = all bishops are equal
- Catholic = pope is vicar of Christ, above all other

DIFFERENT LITURGICAL PRACTICES

- *CATHOLIC*
  - Altar in full view
  - Three-dimensional statues
  - Secular artists
  - Development of realist style
  - Baptism by sprinkling
  - Left-to-right cross
  - Parish priests are celibate
- *ORTHODOX*
  - Icon-screen hides altar
  - Two-dimensional icons
  - Monastic artists
  - Rejection of realist style
  - Baptism by immersion
  - Right-to-left cross
  - Married parish priests

MISSIONARY OUTREACH

- 9th c. = Cyril and Methodius lead mission to Slavs
- 10th c. (988 CE) = conversion of Vladimir of Kiev
- State served as an instrument for the Christianization of the Eastern Slavs
- Church subordinated to state (= Byzantium)
- 1598 = Moscow becomes Patriarchate

WESTERN CHRISTENDOM

- Constantine (4th c.) = capital city of the Roman Empire moved from Rome to Byzantium
- 5th c. Germanic tribes conquered Rome
- Family-based traditions of the Germanic tribes replaced edicts of Roman emperors

FRAGMENTATION OF WESTERN ROMAN EMPIRE (6th c. CE)

- Franks in Gaul
- Visigoths in Spain
- Ostrogoths (later Lombards) in Italy, Austria, and Hungary

CONVERSION TO CHRISTIANITY = CONVERSION TO ISLAM

- Same methods
- War
- Trade
- Education

ARAB CONQUESTS + RISE OF CAROLINGIANS

- 8th c. Muslim Arabs defeated Visigoths in Spain + occupied southern part of France
- Frankish/French Charles Martel (in Latin Carolus) stopped Muslims at Battle of Tours
- Carolingians famous for military effectiveness (fiefs = grant of land in exchange for military support)

FEUDALISM

- Society in which local lord was dominant + offered security in return for allegiance from his dependents or vassals
- System of mutual rights and responsibilities

CHARLEMAGNE (8TH C. CE)

- Charles Martel's grandson
- Empire = Gaul, parts of Germany and Italy
- Jewish traders invited to live in Western Europe
- Palace school = center of a modest renaissance of classical learning

TREATY OF VERDUN (9TH C.)

- Divided empire into three parts:
  - French-speaking in the West
  - German-speaking in the East (Germany)
  - Kingdom of Burgundy
- Why? Germanic tradition of splitting property equally among sons

GERMANIC ORDER IN THE HOLY ROMAN EMPIRE

- Time of invasions (Arabs, Franks, and Vikings)
- Most cities lost population
- Roman roads fell into disuse
- Subsistence economy
- Tendency was to rely on local resources
- Self-sufficient farming estates known as manors became centers of agriculture

SERFDOM

- Serf comes from *servus* (= slave) in Latin
- Belonged to manor + owed dues and obligations
- Serfdom was not slavery (legally, serfs could not be bought or sold as chattels)
- Origins: Lack of central government + villages were more vulnerable + need for protection
- 10th-11th c. most peasants in England, France, and West Germany were unfree serfs

CHURCH-STATE RELATIONS

- Germanic custom law gave supreme power to the king in all matters
- Canon law based on Roman precedent gave Church jurisdiction over all of Western Christendom
- Big question: who will have control over appointments of bishops?
  - Bishops = landowners, vassals, served in military

MONASTICISM

- Groups of celibate monks or nuns living together in organized communities
- Monasteries = centers of learning, conversion, healthcare, deforestation, education, welfare
- Preserved Roman works
- St. Benedict and St. Scholastica (5th-6th c. CE, Italy) developed rules of poverty, celibacy, obedience to the abbot

REVIVAL OF WESTERN EUROPE (11TH C. CE)

- Before 11th c. subsistence economy
- After 11th c.
  - Agrarian surpluses + return of money-based economy
  - This allowed larger numbers of craftspeople, construction workers, and traders

ORIGINS OF GROWTH

- New technologies in agriculture
  - Horse collar for plowing wet lands
  - Plow with a knifelike blade
  - Three-field system
    - Alternation of wheat and rye with oats, barley or legumes
    - Leaving one third uncultivated [before peasants left half the land fallow]
- Warmer climate

SELF-GOVERNING CITIES IN FLANDERS AND ITALY

- Groups of leading citizens demanded privilege of self-government from lay or ecclesiastical lords
- Venice competed with Muslim ports
- Northern Bruges rivaled with Venice

HANSEATIC LEAGUE (13TH-14TH c.)

- New trade zone emerged in northern Europe
- Economic and defensive alliance of towns
- Included Flemish (today's Belgium and Netherlands) + German cities
- Traded with Novgorod (in today's Russia) and London

RISE OF ROYAL AUTHORITY

- Cities and kings had a common interest
  - Depend less on landed nobility
- Struck a deal
  - King granted cities right to form own government and law courts
  - Cities paid taxes to king who could hire professional soldiers to defend realm

NO SOCIAL EQUALITY

- Three estates:
  - "Those who pray (clergy)"
  - "Those who fight (nobles)"

  - "Those who work (peasants)"
- First and second estates enjoyed rights and privileges denied to third estate

CHIVALRY

- Code of ethics and behavior among nobles promoted
  - by church to avoid/control fighting among Christians
  - by women to soften relations between sexes
- Involved special initiation as a knight (the candidate placed his sword upon a church altar and pledged his service to God)

UNIVERSITIES

- Christians used madrasa (Muslim academies) as models
- Two big differences:
  - Universities were degree-granting corporations
  - Universities imparted both religious and nonreligious learning
- International language of science = Latin

UNIVERSITY OF PARIS (12TH C.)

- Organized faculties of arts, law, medicine, and theology
- Granted "licenses" to teach + students became "masters" or "doctors" after defending thesis
- Became center of scholasticism (= system of study combining Christian faith and ancient Greek philosophy—Aristotle)
- Parisian scholars were familiar with Muslim and Jewish scholarship

MOST FAMOUS TEACHER IN PARIS: THOMAS AQUINAS (13TH C.)

- Scholars must use both faith and reason to learn about God and the world
- Condemned antisemitism + forced conversion of Jews

GOTHIC ARCHITECTURE (12TH-14TH c.)

- Pointed arches (they replaced older round Romanesque arch)
- Towering walls
- Sunlight
- Glass windows
- Height surpassed only in the 19th c.

CHRISTIANS AND MUSLIMS IN SPAIN

- Through works of Central Asian-Persian al-Khwarizmi, Europeans acquainted with algebra, Arabic-Hindu numerals, and Egyptian astronomy
- Through the works of Central Asian Ibn Sina (Avicenna), Europeans became acquainted with the Greek and Indian medicine + Greek philosophy
- These works reached Europe through Spain

MUSLIM IMPACT

- Paper

- Fine steel swords
- Water mills and windmills
- Navigational innovations (triangular "lateen" sail + astrolabe for navigating by sun and stars)
- Muslim love poems and songs

CRUSADES (MOTIVES) = RELIGIOUSLY INSPIRED CHRISTIAN MILITARY CAMPAIGNS AGAINST MUSLIMS

- Religious fervor (pilgrimages played an important role in peoples' religious life)
- Younger noble sons could not inherit land. Craved for titles and lands
- Way to stop fighting other Christians + way for kings to keep nobility occupied
- Merchants wanted secure trading posts in Middle East
- Byzantine ruler called for the pope's help (the Muslim Seljuk Turks had invaded Anatolia and threatened Constantinople)

IMPACT OF CRUSADES

- Soap, hospital design, pasta, paper, sugar, cotton cloth, colored glass + Greek manuscripts entered Western Europe
- Words like *cotton, coffee, sugar, tariff, musket, admiral, algebra, zero* all come from Arabic

CLICKER NOTES

DAILY LIFE IN MEDIEVAL TIMES

Please write four facts that you learned from the clips today

1.

2.

3.

4.

HISTORIANS LIKE TO ARGUE: CHIVALRY

1. What is the main thesis that the presenter tries to challenge?

2. How does he prove his point? What is his source base?

3. Are you convinced?

CHILDHOOD IN THE MIDDLE AGES

1. How did previous historians view childhood in medieval times?

2. Do historians agree with this position today?

3. What is their source base?

4. Are you convinced?

ROMANESQUE OR GOTHIC?

© Sibrikov Valery/Shutterstock.com

PICTURE 1
BEAUVAIS CATHEDRAL (FRANCE)

A. Romanesque  B. Gothic

PICTURE 2

A. Romanesque B. Gothic

MIDDLE AGES

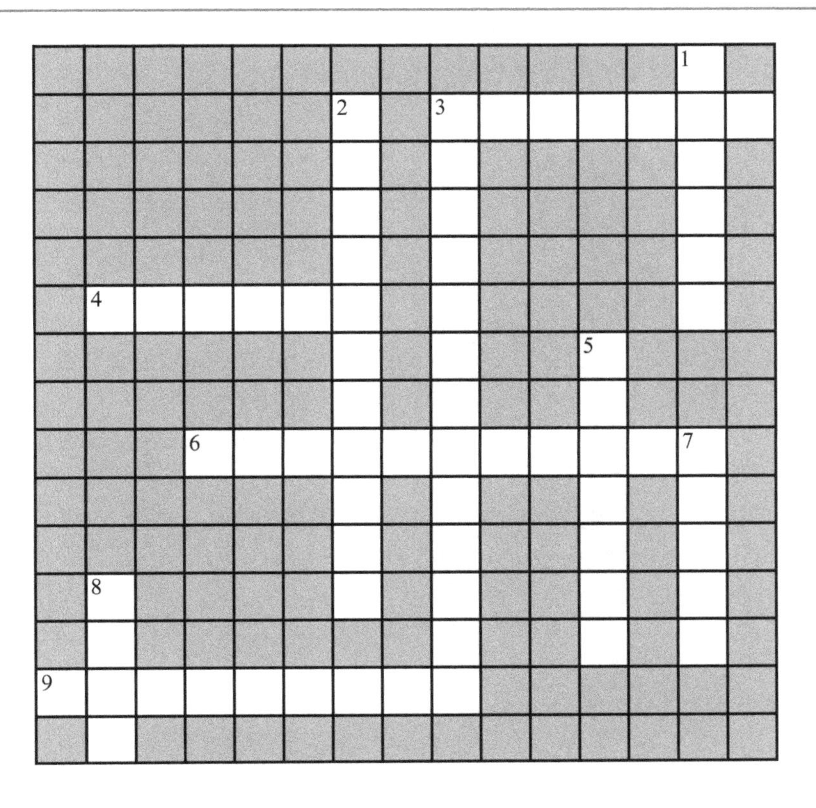

ACROSS

3. A system of labor in which farmers were legally bound to work for and pay fees to a particular landlord in return for protection
4. An individual who holds land given by a lord in return for loyal service
6. A religious movement in which devout men and women withdrew from secular society to live in religious communities
9. A hierarchical system of social and political organization in which an individual gives grants of land to other individuals in return for allegiance and service

DOWN

1. A style of church architecture with pointed arches, towering walls, and stained glass windows (12th century)
2. Frankish king crowned emperor in 800 who patronized learning and invited Jews to settle in Western Europe
3. Method of study based on logic and dialectic. Students had to defend knowledge learned from authoritative texts (Aristotle and Church fathers)
5. Armed mounted warriors whose code of conduct entailed strict devotion to their overlords and to the Church
7. Vast landed estate owned by a noble and worked by peasant farmers; also the main house and castle on the estate
8. Land granted to a vassal in exchange for services, usually military

Chapter 15
The Mongol Empire

© Pierre-Jean Durieu/Shutterstock.com

Two Young Boys Riding in the Middle of the Mongolian Steppe

Genghis Khan on a Mongolian Stamp (ca. 1990)

The Mongolian Steppe

A Mongolian Yurt

MAP OF THE MONGOL EMPIRE (THIRTEENTH CENTURY)

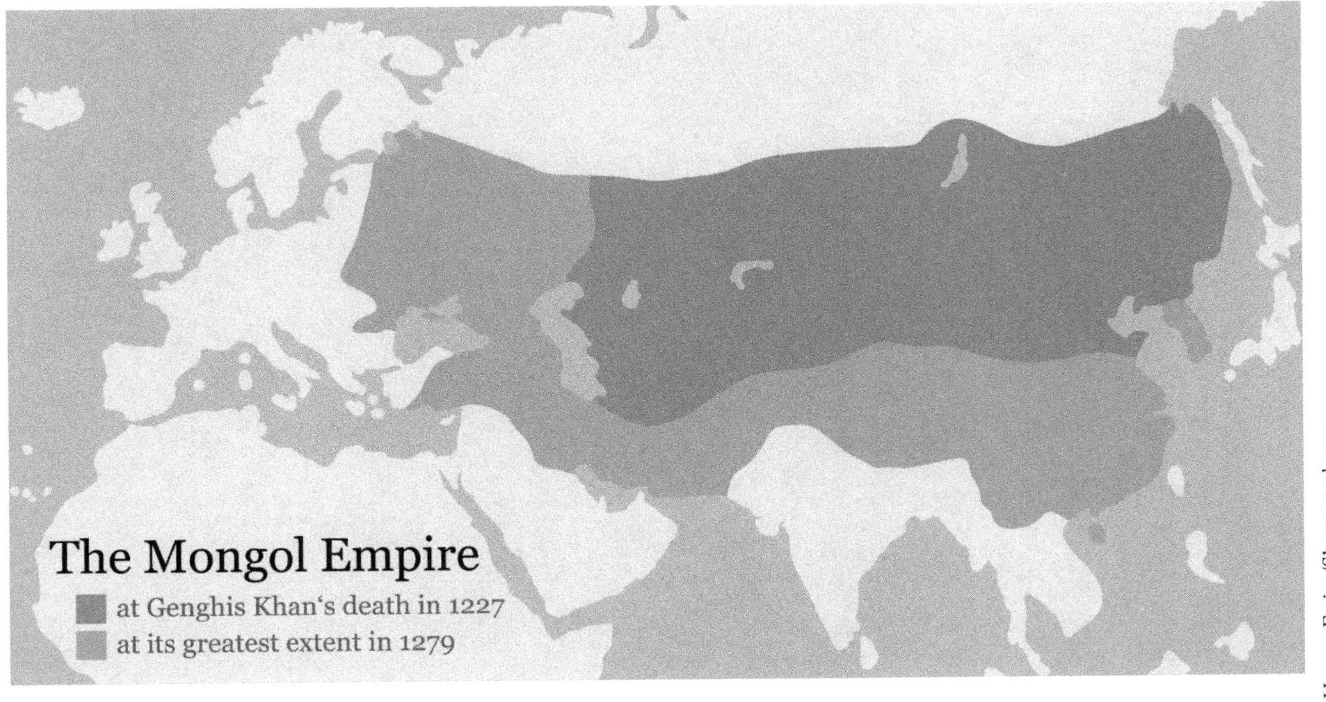

How did ecology impact the expansion of the Mongol empire?

____

____

____

GENGHIS KHAN AND THE MONGOL EMPIRE

MONGOL EMPIRE (13TH-14TH C.)

- Largest contiguous land empire in human history
- Affected four geographical areas
  - China
  - Central Asia
  - Russia
  - The heartlands of Islam: Iran and Iraq
- First time Islamic world under non Muslim power

WHY DID MONGOLS INVADE OUTER EURASIA?

- Climatic changes
- Political infighting forcing losing party to migrate elsewhere

GENGHIS KHAN (13TH C.)

- Temujin, future Genghis Khan, united tribes (Mongol + Turkic speaking)
- Genghis = oceanic or world embracing
- Pastoralist (steppe) community
- Animistic/shamanistic
- Four sons (empire equally divided)

THE FOUR MAIN BRANCHES OF THE MONGOL EMPIRE

- The Yuan dynasty in China (Empire of the Great Khan)
- Jaghatay khanate in Central Asia
- Golden Horde in Southern Russia
- Ilkhanid dynasty (Hulegu's dynasty) in Greater Iran

MORE ON THE FOUR MAIN BRANCHES OF THE MONGOL EMPIRE

- Yuan dynasty in China = learned from Confucian scholars how to govern
- Jaghatay khanate in Central Asia = clang to nomadic ways + conversion to Islam
- Golden Horde in Southern Russia = conversion to Islam + rise of Moscow (Russia ultimately absorbed Golden Horde)
- Ilkhanid dynasty in Iran = Islamization

WHY WERE MONGOLS SUCCESSFUL?

- Extraordinary riders (small, fast horses)
- Superior bows (light, could shoot from great distance)
- Gathered intelligence before attack
- Used terror and intimidation (propaganda)

BORROWED AND IMPROVED NEW TECHNOLOGIES

- Used fire flaming projectiles, catapults, and gunpowder to destroy city wall
- Invented first cannons
- Adapted Turkic-speaking Uyghur alphabet for compiling info
- Learned from Confucian scholars how to govern

CONSEQUENCES OF MONGOL CONQUESTS

- Agriculture disrupted
- Deforestation for grasslands
- In Eastern Central Asia under Jaghatay settled life disappeared (even today)
- Mongols destroyed irrigation works in Persia (*qanat*)
- Cities destroyed (Baghdad, Merv)

RECOVERY FROM MONGOL CONQUESTS

- New cities flourished (Ardabil in Caucasus)
- Bukhara, Samarkand reborn (along silk roads)
- New patterns of commercial activity (trade increased with China + Russia)
- Rice, porcelain reached Middle East and Europe
- Huge demographic shifts = massive immigration of Turkmen to Anatolia and Iran

EFFECTS OF MONGOL CONQUESTS

- New concepts of legitimacy in Middle East
  - Descent from Genghis Khan
  - Later Muslim rulers appealed to such concepts
  - *Yasa* (Mongols' own code of law) versus sharia
- Gunpowder + printing entered Middle East, India, Anatolia (Turkey)
- Stimulated flow of goods + geographical/scientific information (from China to Europe)
- Created infrastructure to travel long distances (extraordinary postal relay service + wagon stations + passport)

PRINTING FROM CARVED WOODEN BLOCKS

- Chinese technology reached the West
- Mongol Ilkhanids in Persia issued paper money in 1294
  - Song dynasty produced first paper money in 12th c. CE
- People in the West started printing holy pictures and card games

SUMMARY

- Mongols changed nature of warfare (first created artillery units and invented very first bronze cannons)
- Printing from carved wooden blocks foreshadowed movable-type printing presses (Gutenberg, 15th c.)
- Scientists and physicians of different ethnic background traveled eastward or westward preparing for the scientific revolution of the 17th century

MONGOLS AND RELIGION

- Originally animists/shamanists
- Mongols showed great tolerance toward religion
- Shiism protected in Sunni Iran
- Churches in Russia did not have to pay taxes

CONVERSION TO ISLAM

- 13th-14th c.—Mongols and Turkic peoples converted to Islam (Golden Horde + Jaghatay + Ilkhanids)
- Sufi missionaries

MONGOLS AND RUSSIA

- Mongol victory = end of Kievan Russia (Rus') + rise of Moscow
- Why? Mongol rule was indirect (Mongols imposed tribute and made Russian princes vassals/tax collectors)
- Khans gave preference to Moscow

MONGOLS AND HEALTH

- Mongol economic integration of Eurasia helped spread diseases (bubonic plague, influenza, typhus, and smallpox) to Europe from Central Asia
- Mongol destruction of dams, irrigation, farmland, and crops exacerbated epidemics

PLAGUE OR BLACK DEATH (14TH C.)

- Fleas responsible for spreading disease from rodents to humans
- China + silk roads through Central Asia + Black Sea + boats carried the disease all around Mediterranean Sea
- In Europe, 30 to 60 percent of population died
- This weakened Mongol Empire (commerce was disrupted) + dynastic strife

MONGOLS KEPT THEIR OWN WAYS

- Mongol women = high status
- Emperor Kubilay often preferred to listen to his Buddhist wife Chabi's advice
- This upset Chinese Confucian administrators

CLICKER NOTES

THE MONGOLS

1. What was Qaraqorum? Where is it located? When was its time of glory?

2. Who was Temujin (Genghis Khan)? Gather as much biographical info as you can from the film.

3. Where do the Mongols come from? Describe their mode of living.

4. What happened in 1204?

5. What happened in 1206?

6. How did Genghis Khan transform the Mongol army?

7. How did Mongol soldiers travel?

8. Which lands did the Mongols conquer? And when?

9. Where was the Khwarizm Shah's empire located? What was its capital? What was its inhabitants' religion?

10. What tactics did Mongols use to win the war?

11. How did the conquests change the Mongol ways of living?

12. What were the consequences of Mongol invasion?

DISCUSS THE IMPACT OF THE MONGOL INVASION ON EUROPE AND THE ISLAMIC WORLD

NEGATIVE IMPACT	POSITIVE IMPACT

THE MONGOLS

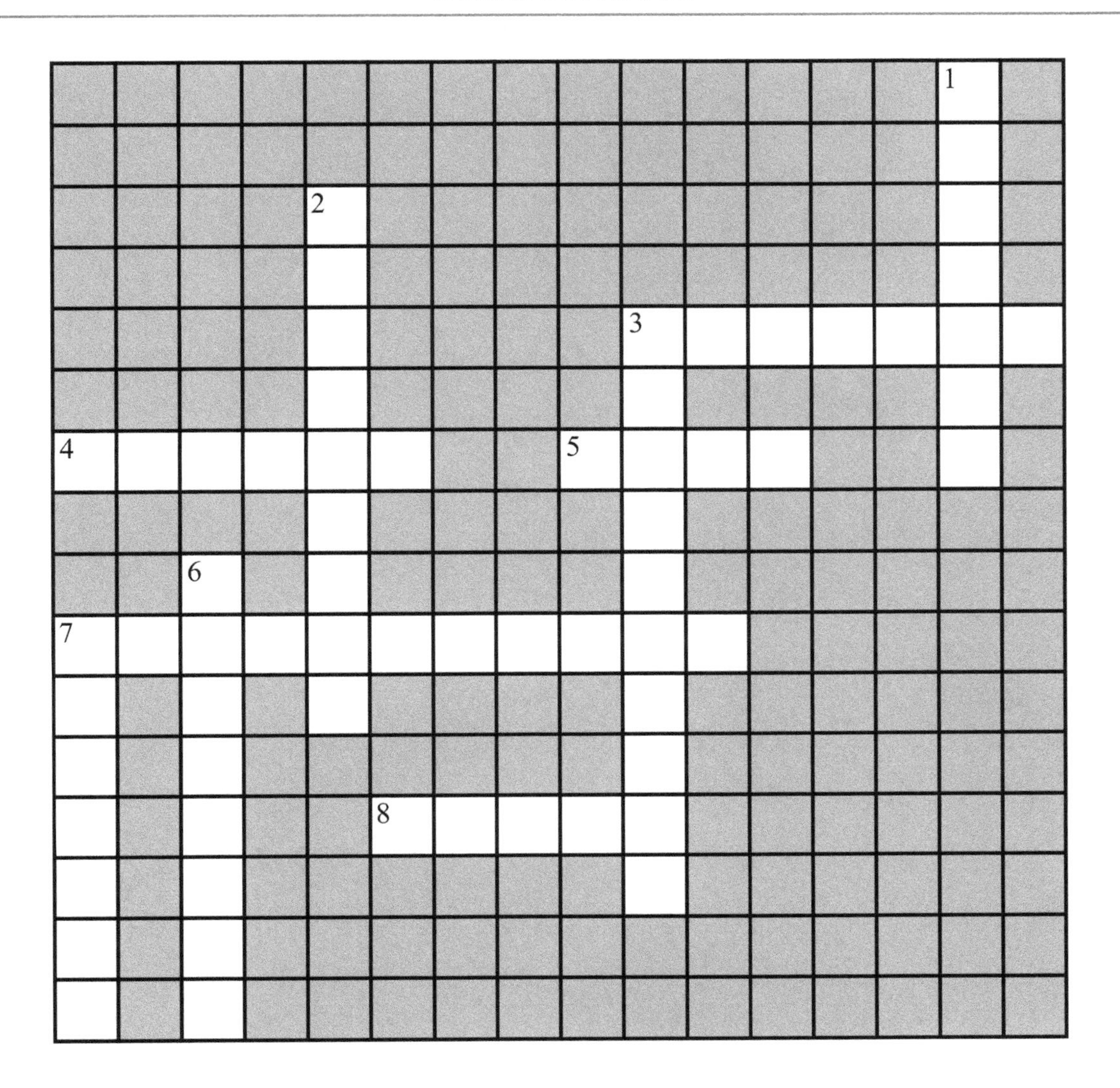

ACROSS

3. This city, capital of the Abbasid empire, was destroyed by the Mongols
4. City in modern-day Russia that rose in the fourteenth century as a consequence of the Mongols' tribute system
5. Mongols' own code of law
7. Name given to the Mongols of Southern Russia from 1240 to 1480
8. Famous underground irrigation system in Persia destroyed by the Mongols

DOWN

1. Vast autonomous region of the Mongol empire, governed by a Khan
2. Mixture of saltpeter, sulfur, and charcoal invented by Daoist priests in China
3. Bubonic plague that killed millions of people in Europe, North Africa, and Asia in the fourteenth century
6. Mongol dynasty in Persia
7. Mongol warrior and conqueror who created the largest land empire in the world

Buddha and Genghis Khan Back in Mongolia

BY JOHN NOBLE WILFORD

ULAN BATOR, Mongolia— In a makeshift temple at the foot of a rocky cliff in the Gobi Desert, four aged lamas in saffron robes sat cross-legged on carpets before a shrine to Buddha. They were fasting, meditating in silence and perusing sacred texts of Tibetan Buddhism.

Outside under a harsh sun lay the crumbling mud-brick ruins of the Ulgii Khiid Monastery, which once held more than a thousand monks. It had been sacked by the repressive Communist Government that ruled Mongolia for 70 years, a Government that destroyed more than 700 monasteries in the 1930's, burning books and executing thousands of lamas.

Now some surviving lamas are returning to occupy Ulgii Khiid and several dozen other monasteries. They live and worship in yurts, the customary dome-shaped dwellings of Mongolian nomads. They are teaching younger men to be monks and making plans to rebuild the temples.

The scene of religious revival amid revolution's ruins illustrates the extraordinary changes under way in Mongolia. In yet another reverberation from the collapsing Soviet empire, the dictatorship tied to Moscow was ousted in March 1990 and replaced with a more reform-mined Communist Government and a Parliament that includes increasingly assertive democratic opposition parties.

With that new political freedom has come a resurgence of Mongolian nationalism. One of its most fundamental expressions is the official and popular support for Buddhism, the traditional religion of Mongolians for centuries.

In an interview in his private reception room at the Gandan Monastery here, the Khama Lama, the country's ranking Buddhist cleric, said, "This movement is very strong, not only among Buddhists but all people."

Other religious leaders, officials, intellectuals and diplomats agreed. Although few Mongolians may have a deep knowledge or belief in Buddhism, they said, anything from the pre-Soviet days is encouraged and celebrated.

That goes also for Genghis Khan, the legendary 13th-century warrior who united the Mongol tribes and set out with the Golden Horde to conquer much of Eurasia. The empire controlled by him and his immediate successors included China and Russia and extended all the way to Hungary.

Since Communists came to power in a revolution here in 1921, the Soviet Union had tried to suppress the name and memory of Genghis Khan, in part because no national hero could be allowed to eclipse Lenin but also because of residual embarrassment over being one of the conqueror's victims. But now Genghis Khan rides again.

Within months of the fall of the Soviet-backed Government last year, Mongolian scholars convened an international conference to commemorate the 750th anniversary of "The Secret History of the Mongols," the epic describing the glory days of Mongolian power and influence.

In a message to the conference, President Punsalmaagiyn Ochirbat endorsed the rehabilitation of the national hero and spoke of the importance of "The Secret History" for "the renewal of the Mongolian society, consolidating of their national integrity and social progress and for the restoration and development of their historical and cultural heritages." Statue of Stalin Removed

At a diplomatic reception a few months ago, Mr. Ochirbat came up to an American military officer and congratulated him on the victory in the Persian Gulf war.

"Our countries now have one thing in common," he said. "We both have conquered Baghdad."

A statue of Stalin in the capital is gone, and billboards bearing Lenin's likeness have been painted over. But the image of Genghis Khan is everywhere, on banners at a trade fair and postcards and in souvenir shops. A vodka is named for him, and a local rock group sings his praises.

A Japanese-Mongolian expedition is spending its second summer in the nearby mountains searching for Genghis Khan's grave. If it is found, a museum will be built at the site.

Dr. Shagdaryn Bira, secretary general of the International Association for Mongol Studies and a member of the Mongolian Academy of Sciences, said the expedition had found several hundred tombs, including some from the 13th century, but not the one they were looking for.

"I'm against making Genghis Khan a demigod," Dr. Bira said. "He was neither a devil nor a god. He was a man."

Even so, Dr. Bira said, efforts were being made to enlist international support for archeological excavations at Khara-Khorum, Genghis Khan's capital. The ruins, west of Ulan Bator, are a regular tourist stop. But research there was suspended years ago, when Mongolians were discouraged from studying their early history, and so the country has a shortage of archeologists.

Officials of the Academy of Sciences, reaching out for contacts with the West, are capitalizing on their paramount scientific asset: dinosaur remains. A group of French and Italian paleontologists arrived in early July to examine the rich fossil beds of the Gobi. Paleontologists from the American Museum of Natural History in New York, working with Mongolians, have begun a planned three-year campaign of fossil hunting. Found Monastery Ruins

It was when the members of the American-Mongolian expedition were crossing the broad, desolate landscape between the towns of Saynshand and Dalan Dzadagad that they found the monastery ruins.

Cresting a dusty ridge, they caught sight of what appeared to be some "lost city" of archeological legend. It was the ruins of a small monastery known as Khonchiyn Khural. On the ground the Americans found spent cartridges and a flattened Buddhist statue with bullet holes, legacies of the sack. The place was utterly abandoned.

Most of those monasteries were destroyed in 1936 in what Renchingiyn Otgon, director of the State Central Library, likened to China's Cultural Revolution in the 1960's. He said more than half of the country's books, many of them priceless hand-printed religious works in Sanskrit and Tibetan, vanished.

In a few more hours of cross-country travel, the expedition came upon Ulgii Khiid, where the white cloth of yurts glistened among the earthen ruins.

The American scientists were given a taste of Mongolian hospitality, tea with camel's milk, and a few were allowed a brief visit inside the temple yurt. The lay leader of the settlement, a young man named Sandag, told them that 12 lamas and several young monks had taken up residence at the ruins late last year. With help from local herdsmen, they hoped to begin rebuilding the temple next year.

The Khama Lama said that some 100 monasteries have been reoccupied in at least a symbolic form, but that their restoration would be slow because of a lack of money and a shortage of monks. Many of them are old -- the Khama Lama is in his 80's -- and only recently have younger men started joining the monasteries.

Michael Sautman, an American businessman here and a practicing Buddhist, expressed concern that "the whole tradition of Tibetan Buddhism, with its scholasticism and deeper meditative arts, have been lost in Mongolia." Many of the lamas, he pointed out, had to spend years outside the temple, living as laborers and herdsmen. Only the Gandan

Monastery in Ulan Bator was permitted to reopen earlier, in 1944, but more as a showpiece than as a vibrant religious center.

Mongolians are adherents of the Gelug sect of Tibet, commonly called the Yellow Hat Buddhists for the yellow caps worn by monks to symbolize their rigid rules on learning, discipline and celibacy, and followers of the Dalai Lama. Indeed, Dalai is a Mongolian word meaning "ocean of wisdom." New Generation to Train

Kushok Bakula, considered a living Buddha and India's Ambassador to Ulan Bator, agreed that it would take years to train a new generation of monks. He and other Buddhist leaders are hoping that a visit from the Dalai Lama would set off a wave of religious fervor, bringing back more people to worship and inspiring the young to take up the monastic life.

But the issue of the Dalai Lama's visit has exposed the limits of religous freedom in Mongolia and served as a reminder that this country of two million people is landlocked between two giants, China and the Soviet Union.

With its withdrawal of most trade, military forces and economic aid, the Soviet Union has left the once-dependent Mongolians to fend for themselves. Food and gasoline shortages are spreading, and Mongolians fear that conditions will worsen. They have made overtures for new trade ties with Japan, South Korea and the West, but they are restricted because their only transportation links are through Soviet and Chinese territory.

Under the circumstances, the Mongolian Government has sought to cultivate relations with China. And since Beijing looks with disfavor on any show of respect for the Dalai Lama, who disputes China's sovereignty over Tibet, cautious Mongolian officials withheld permission for the religious leader's planned visit here this month.

Not that the monks at Gandan are sitting still, waiting for a visit from the Dalai Lama. At noon one day in early July, they were out inspecting the construction of a new temple, young monks lending a hand to the old ones as they leaned over the foundation trenches. One of the few other construction projects in the city is the imposing glass edifice for a luxury hotel to open next year -- the Genghis Khan.

After reading article 9, answer the following question: what is the significance of Genghis Khan and Buddhism for modern Mongolia?

Chapter 16

After the Mongols in the Christian West

Astronomer Nicolaus Copernicus (d. 1543)
What is Copernicus famous for?

Printer Johannes Gutenberg (d. 1468)

Movable Type Printing Press

Original Metal Letter

THE RENAISSANCE

BLACK DEATH (14TH C.) AND ITS CONSEQUENCES

- Workers who survived asked for higher pay
- When landowners refused, this triggered major peasant revolts
- Serfdom practically disappeared in Western Europe
- Landowners had to adapt (in England they pastured sheep for wool, others grew crops that necessitated less care)

POGROMS

- = violent attacks on a minority community characterized by massacre and destruction of property
- Jews fled eastward to Poland and Russia (Ashkenaz)
- No pogroms in Middle East. At the time of the Reconquista, Jews (Sepharads) found refuge in Ottoman Empire

"INDUSTRIAL REVOLUTION" IN MEDIEVAL EUROPE

- Mining, metalworking, and craft mechanization expanded
- Mills powered by water or wind ground grain, sawed logs, crushed olives
- Designs of these mills dated back to the Greeks and Romans and reached Europe through Islamic Spain
- Techniques of deep mining developed
- Building boom (stone quarrying in France increased)

ECOLOGICAL CONSEQUENCES

- Quarries and mines scarred the landscape
- Dams and canals altered the flow of rivers
- Dense forest disappeared (a single iron furnace consumed a lot of trees)
- Pollution: 14th c. = first recorded anti-pollution act in England (urban tanneries and human waste polluted rivers)

URBAN REVIVAL

- Until end of 14th c., cities in Islamic world were the largest
- In the 14th c. port cities and cities located on rivers rivaled Muslim cities in size (Venice, Genoa, Hanseatic League of Cities)
- European cities borrowed a lot from the Muslim world (papermaking, glassblowing, ceramics, and sugar refining)

CIVIC LIFE

- Italian, Flemish, and German cities were independent states
- Other cities in Europe held royal charters exempting them from the authority of local nobles
  - If a person found a job in the city, this person could claim freedom
  - This promoted social mobility

GUILDS

- Associations of men (rarely women), such as artisans, merchants, or professors, who worked in a particular trade
- Promoted their economic and political interests
- Guilds set prices + trained apprentices
- Found also in Islamic World

DRAWBACKS

- Denied membership to Jews
- Protected interests of families that already belonged to them
- Perpetuated male dominance of most skilled jobs
  - Still, women in a few places managed to join guilds on their own as the wives, widows, or daughters of male guild members

BANKS

- Emergence of wealthy merchant-bankers in the 15th c.
  - Money changing + loans + investments + checking accounts + credit
  - Organized private shareholding companies
  - Even gathered information to invest properly
- Most famous family in Italy: the Medici (banks in Italy, Flanders, London)

LATIN CHRISTIANITY AND MONEYLENDING

- Moneylending was sinful
- For this reason Jews predominated in this area
- Another possibility was to repay loans in a different currency at a rate favorable to the lender or to offer a "gift"

RENAISSANCE

- Began in Italy, Florence (wealthy banking city-state)
- Goal = restore culture of classical Rome in Italy
- Roots of the Renaissance:
  - Rediscovery of ancient Greek and Roman art and thought preserved in Islamic and Byzantine scholarships
  - Funded through commerce with the East + banking system that supported this trade
  - Enhanced by new technologies (paper and printing developed in China and spread by Muslim traders)

EDUCATION

- Development of a humanistic curriculum in secondary schools of Europe
  - Humanism = scholarly study of the Latin and Greek classics and the ancient Church fathers to promote a rebirth of ancient norms and values
- Humanists advocated a liberal arts program
- Emphasis on grammar, rhetoric, poetry, history, politics, and moral philosophy

RELIGION

- Religious themes were still prevalent BUT the emphasis was more on life in this world than the afterlife
- Rediscovery of Greek and Latin led to new translations of Bible = Erasmus, Dutch scholar of Rotterdam (corrected mistakes in earlier Latin translations in 15th-16th c.)
- Renaissance painters and sculptors portrayed
  - Ancient deities and myths
  - Daily life or depictions of grief and love

ARTISTIC REALISM

- Painter Giotto (13th-14th c.): his portrayal of saints and biblical scenes pioneered artistic realism
- Lifelike portraits (not stiff, staring portraits as in Byzantine art, aimed at inspiring awe)
- Depth + perspective (earlier medieval art was more flat and stylized)

LEONARDO DE VINCI : RENAISSANCE MAN

- Military engineer, self-taught botanist, painter
- Dissected corpses to learn anatomy
- Used oil paints + tried to represent inner moods
- Worked in many media (bronze sculptures, large frescoes)
- His notebooks contain imaginative designs for airplanes, submarines, tanks
- Art for him was a way of examining nature more closely

USE OF VERNACULAR LANGUAGES

- Latin texts translated into Italian
- Pioneer of vernacular literature: Dante, *The Divine Comedy*
  - Religious overtone
  - But also bitter critique of the mores of the church and rulers
- Francesco Petrarch
  - Revived literature of ancient Rome
  - Wrote beautiful love sonnets in Italian to a woman who died in the plague
- Giovanni Boccaccio
  - Described people as they were (not as they should be)

RENAISSANCE (DIFFERENCES)

ITALY

- Classical Greek and Roman past was important
- Roman ruins

NORTHERN EUROPE

- Classical Greek and Roman past was not important
- Focus was on daily life

JAN VAN EYCK IN FLANDERS (NOW IN BELGIUM) (14TH-15TH C.)

- Mastered the use of oil painting
- Realistic (meticulously portrayed the appearances of his subjects)
- Depiction of daily life (domestic themes)

GERMAN JOHANNES GUTENBERG (15TH C.)

- Began a communication revolution
- Began using movable type to print books
  - Movable type = small metal blocks for each letter arranged in a frame to print one page. They could be rearranged to print others
  - Before this, texts were copied by hand on animal skin parchments

GERMAN ALBRECHT DURER (15TH-16TH C.)

- Best known for woodcuts (carvings on wooden blocks) and engraving (etchings in copper)
- This technique allowed the multiple copies of the same work
  - Drawings could be used to illustrate books
  - This made work available to ordinary people (not just the wealthy)

POLISH GERMAN: NICOLAUS COPERNICUS

- Confirmed that the Earth orbited around the Sun (on the basis of Greek Ptolemy's research and Islamic scholarship)
- Such discovery challenged the Church's assumption that the Earth was at the center of the universe

CLICKER NOTES

WHAT MAKES THE NORTHERN EUROPEAN RENAISSANCE UNIQUE?

1.

2.

3.

4.

RENAISSANCE

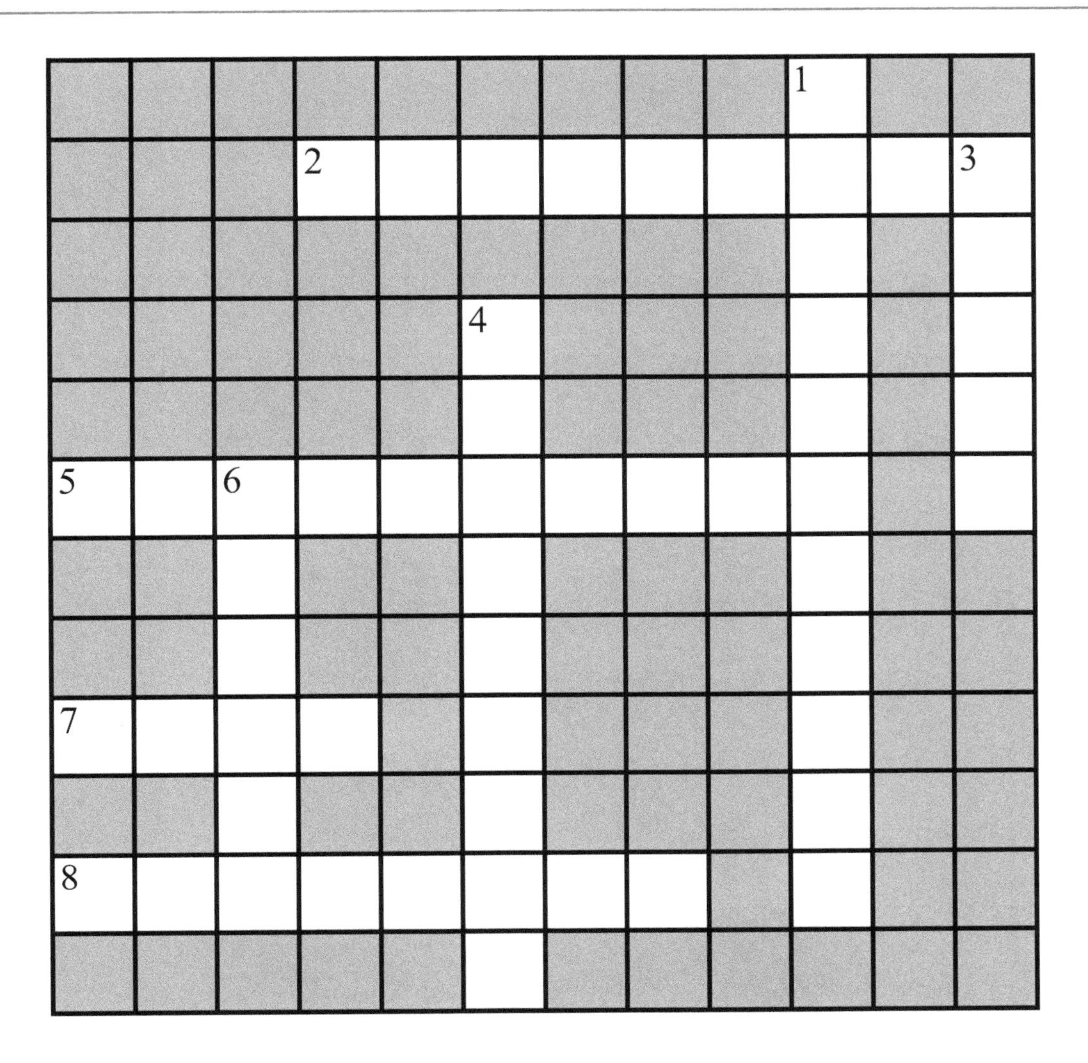

ACROSS

2. German inventor of movable metal type. His innovation led to the publication of the first printed books in Europe.
5. Polish-German astronomer who proposed that the earth and other planets orbit the sun
7. Agricultural laborer legally bound to a lord's property and obligated to perform services for the lord in exchange for protection
8. An intellectual movement in Renaissance Italy based on the study of the Greek and Roman classics

DOWN

1. Literally "Rebirth" of classical culture or period of intense artistic and intellectual activity that started in Italy
3. Association of a particular occupational group formed to promote its commercial and professional interests
4. The _______ League was an economic and defensive alliance of free towns in northern Europe founded in the 13th c. but most powerful in the 14th c.
6. Violent attack on a minority community characterized by massacre and destruction of property

Article Ten

Springtime of the Renaissance: At the Dawn of Magnificence

A new exhibition celebrates the masters of the early 15th century

ITALY'S self-esteem is at a low ebb. Last month's election was won by clowns. The Italian economy is struggling and the new pope is reading the results of an unhappy inquiry into the Roman Catholic church bequeathed to him by his predecessor. But for visitors to Florence, the dome that Filippo Brunelleschi built in 1436 rises "above the skies, ample to cover with its shadow all the Tuscan people", according to a contemporary commentator.

Brunelleschi's dome is a reminder that Florence was once the fulcrum of a new world. It also symbolises the intellectual and artistic movement that came to be known as the Renaissance, which many argue had its birth here.

A new exhibition devoted to this pivotal moment explores how a small group of remarkable men in Republican Florence broke through the carapace of medieval thought to rediscover ancient learning. They unleashed a wave of study and creativity that spread across politics, art, architecture, literature and philosophy, and has become a byword for renewal.

The movement placed man, rather than God, at the centre of the universe. Nature, more than heaven, was to be the proper subject for an artist. The causes and consequences of this shift have been endlessly debated. But the curators of this show argue that it was in sculpture, not painting, that the Renaissance was really born.

They suggest that this aesthetic revolution first showed itself in two bronze relief panels made by Brunelleschi and his compatriot, Lorenzo Ghiberti, in 1401 in a competition to create a set of new bronze doors for the cathedral's neighbouring baptistery. Both artists sculpted scenes of vivid human drama drawn from the story of Abraham's sacrifice of his son, Isaac (pictured). Although Ghiberti eventually won the competition, both artists combined a Gothic elegance of costume and scenery with realistically modelled figures that owe a visible debt to classical sculpture.

A collaboration between the National Museum of the Bargello, the main sculpture museum in Florence, and the Louvre in Paris, the exhibition has brought together 137 objects, from museums and private collections all over Europe and America. Most of the curators' requests were granted, with the result that almost all the great pieces of this Renaissance jigsaw are here.

The first few rooms contain examples of the Roman sculptures that provided the stimulus for the innovative Renaissance pieces placed beside them. The magnificent Roman vase known as the Talento Crater, which once stood outside Pisa cathedral, inspired Nicola Pisano, a 13th-century master sculptor who was an important precursor to the Renaissance. Roman portrait-busts have been placed nearby, alongside the monumental statues of prophets and saints that they influenced.

The most important artist exhibited here is Donatello, who started out as an assistant to Ghiberti and Brunelleschi. Brought together, on his home ground, are some of Donatello's finest sculptures, such as the 1425 hollow bronze statue of St Louis of Toulouse, which has been restored specially for this exhibition. The show enables the visitor to trace directly Donatello's decisive impact on his contemporaries, including, among others, his longtime collaborator Michelozzo, whose work is represented here by two marble angels from London's Victoria and Albert Museum.

The exhibition also points out how ideas expressed first in sculpture became the catalyst for new ideas about beauty and truth in painting. One example is Donatello's delicate use of precise central-point perspective when carving in very shallow relief, which was first seen in 1417, in a scene of St George fighting the dragon. This development was immediately seized on by the painters of the day—Masaccio, Filippo Lippi, Paolo Uccello and Andrea del Castagno—who wanted to emulate the liveliness of sculpture by creating convincing three-dimensional space on flat canvas and wood.

The Strozzi's thematic approach reinforces the importance of the contribution that sculpture made to the Florentine Renaissance. The result is a selective rather than exhaustive presentation. One room explores how the transformation of Roman *putti* into Donatello's charming childlike spirits represents a concerted effort to reconcile Roman imagery with Christian ideals. Another looks at Donatello's reinvention of the classical equestrian portrait. A whole room is devoted to paintings and sculptures of the Madonna and child. The pieces reinforce the idea that the Renaissance's love affair with antiquity was not at the expense of Christian piety, proving at the same time just how quickly new ideals of beauty could spread.

The exhibition also shows sculpture following Florence's political shift from a citizens' republic to a state that was controlled more and more by powerful families like the Medici and the Strozzi. Donatello's early monumental sculptures were commissioned by guilds or the church. The later portrait-busts of Florentine worthies in Roman garb by Mino da Fiesole reflect a new world of magnificent private patronage.

The elegant geometries of the Strozzi palace are illustrated by Giuliano da Sangallo's wooden model of 1489, placed at the end of the show. As with every exhibition here, departing visitors are expected to spill into the city and seek out the Renaissance elsewhere—in other museums and churches, and under Brunelleschi's dome itself.

"The Springtime of the Renaissance: Sculpture and the Arts in Florence, 1400-1460" is at Palazzo Strozzi in Florence until August 18th before moving to the Louvre in Paris from September 23rd until January 6th 2014

In article 10, what is the main argument of the exhibit?

Chapter 17

After the Mongols in the Islamic World

Mud Brick Mosque in Timbuktu (Mali, Africa)

Part 1: Mali in Africa (Major Islamic Center)

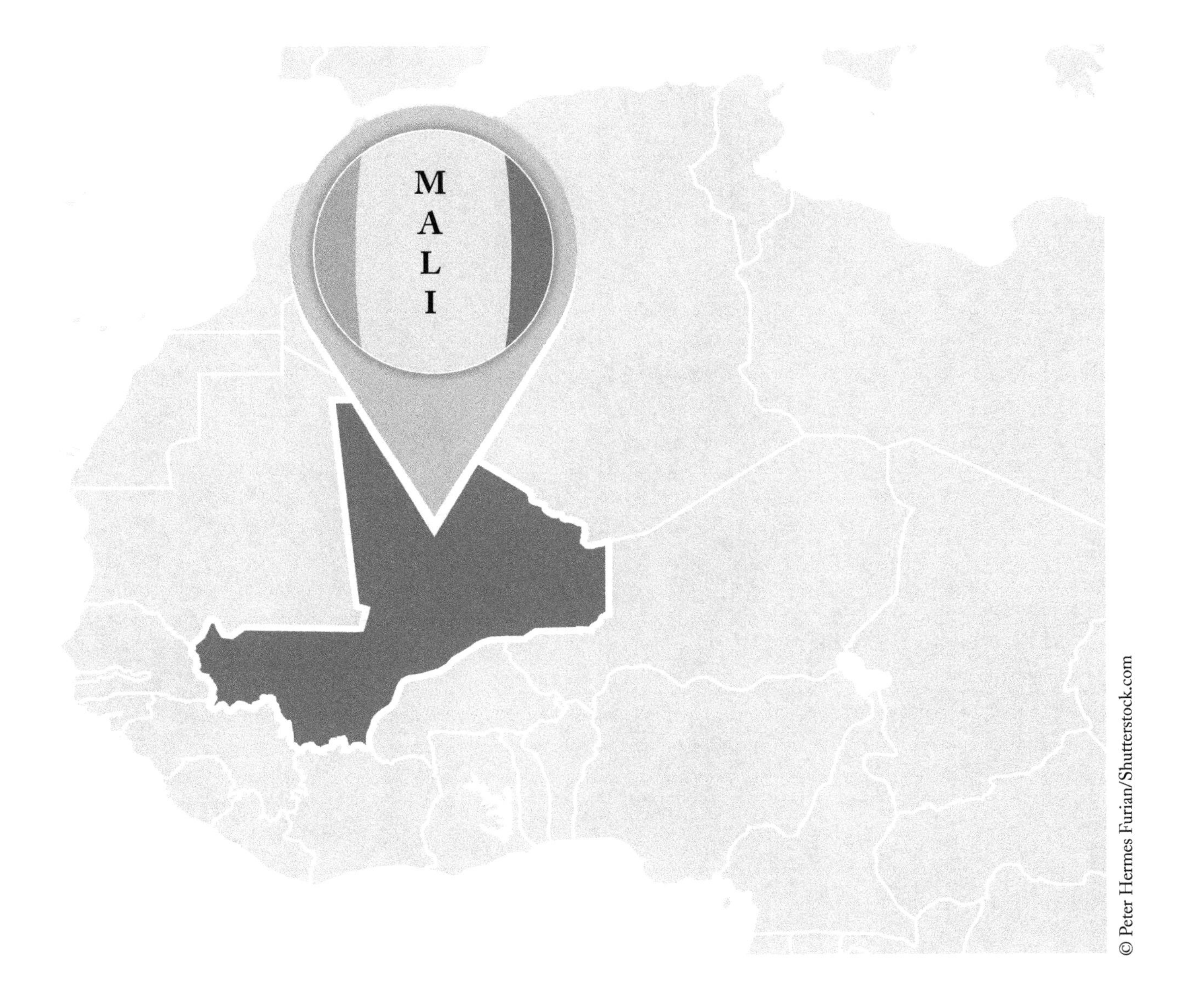

MAP OF MALI TODAY

Instructions: First, find and circle Timbuktu (Tombouctou) on the map. Second, add the name of the main river that crosses Mali.

MALI IN AFRICA

DESTRUCTION OF BAGHDAD BY MONGOLS: WAS THE ISLAMIC WORLD IN DECLINE?

- NO, ON THE CONTRARY
- Islam continued to spread, even at a faster rate
- Mongols and other nomads converted to Islam
- Ottoman Turks extended rule to the Balkans in Europe
- Most of the expansion occurred in
  - Africa
  - South Asia (India)
  - Southeast Asia (Malacca)
- Conquests + trade + Sufism

ISLAM AND NATIVE AFRICAN LANGUAGES

- Family Sufi lineages helped spread Islamic and Arabic culture
- Religious texts were translated into Swahili, Wolof, Hausa, among others
- African languages borrowed a lot from Arabic (= Latin words in Western languages)
- Same phenomenon in other parts of the Islamic world

TIMBUKTU

- Major intellectual center in 15th-16th c. (read article 11)
- Legal opinions could be influenced by commercial interests
- Traded amulets + tobacco + slaves

Article Eleven

When Timbuktu Was the Paris of Islamic Intellectuals in Africa

BY LILA AZAM ZANGANEH

In popular imagination, the word Timbuktu is a trip of three syllables to the ends of the earth. Today this West African city is a slumbering and decrepit citadel at the southern edge of the Sahara, in Mali, one of the poorest countries in the world.

Yet it is here that some of the most astonishing developments in African intellectual history have been occurring. In recent years, thousands of medieval manuscripts that include poetry by women, legal reflections and innovative scientific treatises have come to light, reshaping ideas about African and Islamic civilizations. Yet even as this cache is being discovered, it is in danger of disappearing, as sand and other grit are abrading many of the aging texts, causing them to disintegrate.

"The manuscripts reveal that black Africa had literacy and intellectualism -- thus going beyond the mere notion of Africa as a continent of 'song and dance,' " John O. Hunwick, a scholar who has uncovered some of the writings, said in a recent interview.

Although this rich intellectual heritage is familiar to numerous Africans, many Westerners still believe that Africa had only an oral, nonliterate culture. Comments like those made by the British historian Hugh Trevor-Roper in 1963 still resonate: "Perhaps in the future, there will be some African history to teach. But at present there is none. There is only the history of Europeans in Africa. The rest is darkness."

In reality, Timbuktu was once a haven of high literacy. These manuscripts, some dating to the 14th century and written mostly in Arabic, show that medieval Timbuktu was a religious and cultural hub as well as a commercial crossroads on the trans-Saharan caravan route. Situated at the strategic point where the Sahara touches on the River Niger, it was the gateway for African goods bound for the merchants of the Mediterranean, the courts of Europe and the larger Islamic world.

When the Renaissance was barely stirring in Europe, Timbuktu was already the center of a prolific written tradition. By the end of the 15th century, Timbuktu's 50,000 residents thrived on the commerce of gold, salt and slaves, and hundreds of students and scholars convened at the city's Sankoré mosque. There were countless Koranic schools and as many as 80 large private libraries. Wandering scholars were drawn to Timbuktu's manuscripts all the way from North Africa, Arabia and even Persia.

The bulk of these texts have remained buried for years in Timbuktu's mud homes. Many owners are the descendants of the skilled craftsman class, and the manuscripts often represent a family heritage passed on from generation to generation.

Mr. Hunwick, a professor of history and religion at Northwestern University who has spent 40 years doing research on Africa, came across piles of manuscripts in the musty trunks of a family library in 1999. They were part of a private collection of several thousand manuscripts, some more than 600 years old. While most were written in Arabic, others used Arabic letters to transcribe local tongues like Fulani and Songhay. Mr. Hunwick said he was awe-struck.

The collection was in the possession of descendants of Mahmoud Kati, a 16th-century scholar who, along with others, jotted intricate notes in the margins of his books. Occasionally Kati commented on the texts, but mostly his notes strayed to other topics, from weddings and funerals to floods and droughts. Of a meteor shower in August 1583, he wrote: "In the year 991 in God's month of Rajab the Goodly, after half the night had passed stars flew around the sky as if fire had been kindled in the whole sky -- east, west, north and south. It became a mighty flame lighting up the earth, and people were extremely disturbed about that. It continued until after dawn."

As early as 1967, Unesco recommended the creation of a manuscript conservation center in Timbuktu. Six years later, with financing from Kuwait, the Malian government opened the Ahmed Baba Center in the city, and it has been collecting manuscripts, acquiring more than 18,000 works so far.

"These amount to about 10 to 15 percent of the written potential in Timbuktu and its region," said Ali Ould Sidi, the chief of the city's small but active cultural affairs office. Some scholars believe there are up to one million manuscripts in Mali, about 100,000 of which are in the Timbuktu region. These texts -- possibly the most ancient to survive in sub-Saharan Africa -- offer a window into the ways black Muslim scholars thought and imagined the world around them over centuries.

Unesco designated Timbuktu as a "world patrimony" site in 1989, and the city has since received numerous conservation grants from American foundations and from the governments of Norway, South Africa and Luxemburg. After finding the manuscripts of the Kati collection, Mr. Hunwick became involved in an international effort to preserve and disseminate Timbuktu's written history, in the process creating the Institute for the Study of Islamic Thought in Africa at Northwestern. Closer to home, last year the New Partnership for Africa's Development, based in Johannesburg, has announced plans for a multi-million-dollar Timbuktu initiative to benefit the manuscripts.

There are formidable obstacles, nonetheless. The texts are rotting inside their metal cases. While turning them over to experts might help preserve them, owners are extremely reluctant to let them go, since they represent personal family legacies.

Those that make it out of family trunks have other problems. Human handling by researchers and visitors, as well as a robust black market, are further chipping away at this historical trove. Chris Murphy, a Near East specialist at the Library of Congress who was a co-curator of an exhibition of Timbuktu manuscripts last summer, said in an interview that trafficking was now common practice. "Poverty is such that you can buy these for $2 to $5," he said. "Then they are taken to Switzerland, often, where their provenance will be forged. And they get moved to auction houses where they will be sold for up to $1,000. Sometimes, they can even reach five figures." Often unaware of their bogus provenance, oil sheiks and university collections alike become potential clients.

Sean O'Fahey, a professor of Islamic African history at the University of Bergen, in Norway, who has worked extensively with European aid agencies across Africa, said that part of the problem in Mali, unlike such other African countries as the Sudan, is that most of the people who own these manuscripts cannot read them because they do not know Arabic. "So what you've got in Mali," he explained, "is a kind of break in the intellectual heritage." This gaping rift between past and present, he said, may prove to be the greatest obstacle to preserving Timbuktu's cultural legacy

What did you learn about Timbuktu in article 11?

Part 2: MIDDLE EAST

An Ottoman Janissary (1895)

MEHMET THE CONQUEROR (1432-1481) ON A BANKNOTE FROM TURKEY

Significance: What was the major achievement of Mehmet Fatih?

____

____

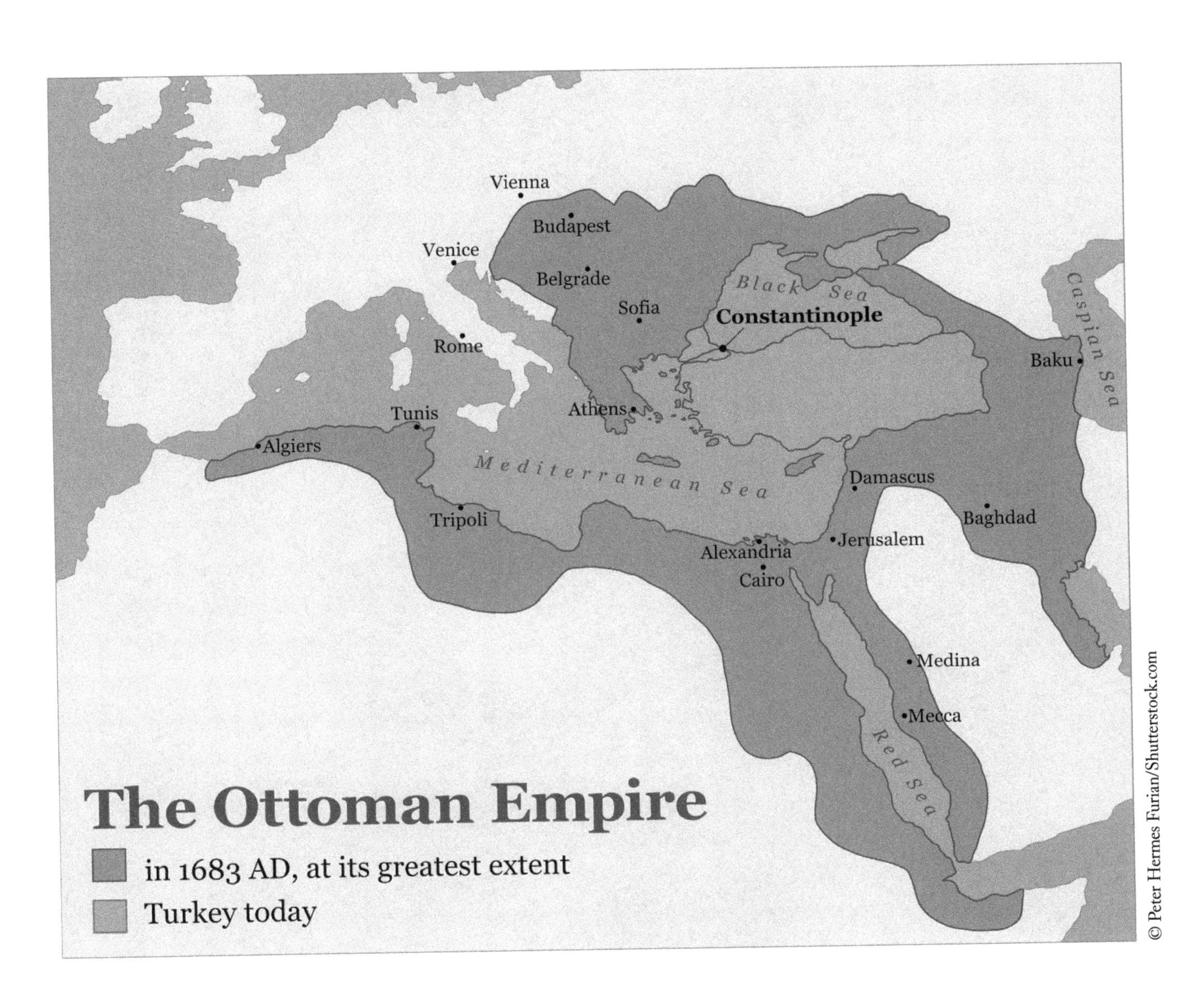

List the countries or regions under Ottoman rule in the seventeenth century.

____

____

____

MIDDLE EAST

OTTOMAN EMPIRE

- Mid-13th c. Ertogrul, a Turcoman chief, established control of NW Anatolia. Paid tribute to the Mongols
- His son, Osman (13th c.), gave his name to the dynasty
- 1453, Mehmet the Conqueror took Constantinople (renamed Istanbul)
- Consequences for Europe:
  - Many Byzantine scholars fled Byzantium for the West, which had an impact on the Renaissance
  - Muslims were at the door of the Christian West
- Suleyman the Magnificent (or the Law Giver) (1495-1566)

WHAT EXPLAINED THE OTTOMAN MILITARY SUCCESS?

- Christians in the Balkans fought among themselves
- 14th c. = black death
- There were competing popes in France and Italy
- In the Balkans, Christian dynastic families looked for Ottoman support

METHODS OF CONQUEST

- Defeated Christian kings could keep their lands
- But they had to contribute troops for Ottoman wars
- The Greek clergy was exempted from taxes and were left in charge of church administration

MILLET

- Nation/religious community system
- Enjoyed limited autonomy
- Leaders collected taxes due to the dynasty + ruled over their communities
- Greek Orthodox viewed Ottomans as their liberators from Roman Catholic rule

HOW DID THE OTTOMANS MANAGE TO BREAK THE CYCLE OF SHORT LIFE OF EARLIER NOMADIC EMPIRES?

- Fratricide
- They were able to settle down Turcomans (Turkish/Turkic nomads)
- As landlords, Turcomans ruled local areas and sent taxes to center or provided troops
- Nomads were pushed to frontier to expand empire

WESTERN EUROPE VS. MIDDLE EAST (MAIN DIFFERENCES)

EUROPE

- Individual ownership
- Hereditary property + landed aristocracy
- Basis of authority: birth
- Power of kings fenced in by nobles' prerogatives
- No more slavery

OTTOMANS

- All lands belonged to the state (Sultan)
- Conquered land given as land grants in exchange of military service (even to Christian princes) but non hereditary (no landed aristocracy)
- Basis for authority = more often merit and ability (as in nomadic societies)
- Slave army dependent on sultan (*devshirme* + Janissaries)
  - Slaves could rise to become leaders of the empire

JANISSARY INFANTRY CORPS

- Composed of Christian-born converts to Islam
- Deshirme = tax [male children in the Balkans were taken away from their parents and given the best education in the empire]
- Slave army (rights of slaves guaranteed by the religious law of Islam)

MAIN CLICHÉS ABOUT OTTOMAN EMPIRE

- Oriental despotism
- Brutal sultans
- Mainly warriors
- Rulers' whim
- No originality (combination of Byzantine, Persian, and Islamic elements)

ABSOLUTE RULERS? YES, MAYBE

- The Sultan had the absolute right to make laws of administration and state
- However, sultans were Muslim believers and were bound to respect the laws of Islam
- Sultans had absolute authority over appointments

ABSOLUTE RULERS? NO (COUNTERWEIGHTS)

- Ulama
- Janissaries
- Turcomans
- Sufis

THE OTTOMAN EMPIRE

1. Who was Suleyman the Magnificent?

2. Who were the Ottoman Turks?

3. Who founded the Ottoman dynasty? What was his dream? What did it represent?

4. What is a gazi?

5. Why did the Ottomans expand to the west?

6. What was the traditional Christian power that the Ottomans attacked? Describe this empire in the 13th century.

7. When did the Ottomans take Bursa? What was its significance?

8. How did the Ottomans manage their vast empire?

9. Why did the Ottomans fear other Muslim noble families? How did this affect the organization of the army?

10. What was the devshirme?

11. Where did the recruits for the devshirme come from? What were they taught?

12. What were the advantages of the *devshirme* for the Ottomans?

13. Who were the janissaries?

14. By the 15th century, where had the Ottoman empire spread? Which city had eluded its grasp?

15. How did the Ottomans prevent the fragmentation of the empire?

16. What was the engineering challenge of taking Constantinople?

17. How was Constantinople conquered? When? By whom?

18. Why didn't the West come to the rescue of Constantinople?

19. What is Hagia Sophia?

20. Who was the greatest sultan of the Ottoman Empire? When did he take the throne?

21. What was his first campaign?

22. How did he capture Rhodes?

23. What was the epithet given to Suleyman?

Part 3: INDIA

© Mario Savoia/Shutterstock.com

Entrance to the Tomb of Akbar, the Great Mughal Emperor (Sikandra, Suburb of Agra, India)

DIWAN-I KHAS, AKBAR'S HALL OF PRIVATE AUDIENCE IN FATEHPUR SIKRI (UTTAR PRADESH, INDIA)

Significance: Who was Akbar and what was taking place in his Hall of Private Audience?

___

___

___

ISLAM IN INDIA

- Islam enters India (10th c.)
- Delhi Sultanate (late 13th c.)
- Mughals (16th c.)

DELHI SULTANATE, 1206-1388

- Late 10th c. Muslim warriors invade northern India
- 1206 = established Delhi Sultanate
- Conquerors initially persecuted Hindus + Buddhists
- Eventually became more tolerant (evidence of state support for Hindu temples)
- Tax exemptions to both Muslim and Hindu religious leaders
- Hindus included within ruling strata

PERSIAN IDEALS OF KINGSHIP

- Strong monarch + monarch's duty to rule justly
- Persian model became more significant after Mongol conquests
- Why? Many Persian artists, poets, and religious scholars fled to India

MUGHAL EMPIRE

- Gunpowder empire
- Founded by leader Babur (1483-1530)
  - Descendant of Tamerlane (Timur)
  - 1523 Conquered northern India

AKBAR, r. 1556-1605

- Greatest Mughal ruler
- Extended imperial territory
- Formed alliances with Hindu princes
  - Married Hindu princesses

AKBAR'S RELIGIOUS REFORMS

- Tolerant toward Hindus and Hinduism
- Revoked laws that discriminated against Hindus
  - Eliminated *jizyah* tax
- Abolished death penalty for apostasy
- Sponsored religious debates
- Created new religious synthesis—"Divine Faith"

AKBAR AND SUFISM

- Akbar favored Sufism
- Akbar's Divine Faith was a kind of *tariqa* (Sufi order) with himself as *pir* (= *shaykh*)
  - Divine Faith was a syncretic monotheism
  - Divine Faith ultimately failed

THE MUGHAL EMPIRE

1. What was the challenge that Islam faced in what is now India and Pakistan?

2. Who brought Islam to India?

3. The film opposes two Mughal rulers: Akbar (r. 1555-1604) and Aurangzeb (1618-1707). In what way are they different? Did they relate to the Chisti Sufi shrines and shaykhs in the same way? What can you say about Akbar's architectural monuments? Aurangzeb's architectural monuments? What kind of Islam did each ruler promote? How did each ruler relate to other faiths?

4. How do Pakistanis and Indians look at these two rulers?

ISLAMIC CIVILIZATION

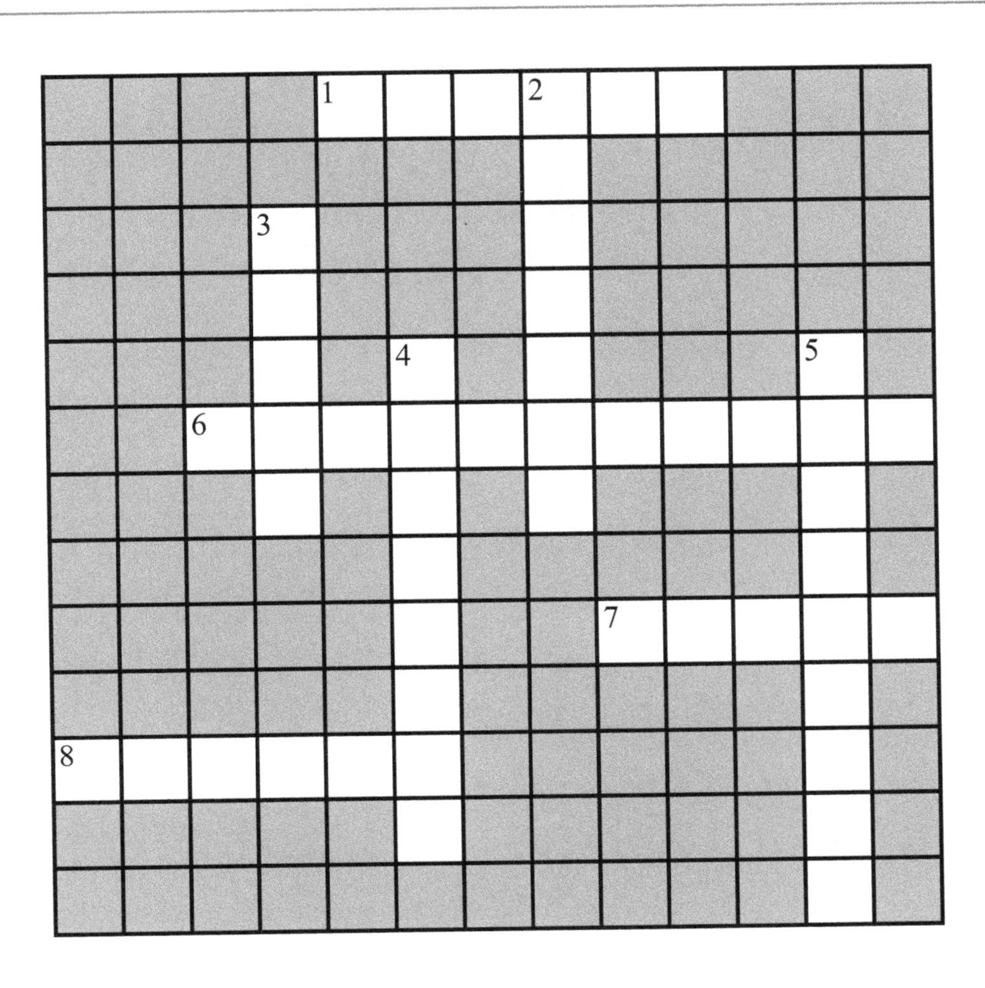

ACROSS

1. The Ottoman Sultan, ______________ Fatih (The Conqueror), took Constantinople in 1453
6. Ottoman infantry composed of slaves who were Christian-born converts to Islam (plural)
7. The ______________ Sultanate preceded Mughal rule in India
8. Ethnic community in the Ottoman Empire that administered its own educational, charitable, and judicial affairs

DOWN

2. Center of higher learning mainly dedicated to the study of Islamic and legal subjects
3. Most famous sultan of the Mughal empire in India. He was known for his great tolerance and pursued a policy of conciliation with Hindus
4. City on the Niger River in Mali. It was a major center of trade and Islamic learning
5. Levy of Christian boys who were raised as Muslims and became soldiers and administrators of the Ottoman Empire

CLICKER NOTES

Chapter 18

The Maritime Revolution and Global Exploration

Santa Maria, Nina, *and* Pinta *of Christopher Columbus*

© Aleksandra H. Kossowska/Shutterstock.com

The Ancient City of Machu Picchu, Sacred Inca Religious Site, in the Peruvian Andes (15th c.)

© ostill/Shutterstock.com

Statue of the Inca Ruler Pachacuti (d. 1471) in Plaza de Armas of Cuzco (Peru)

© steve estvanik/Shutterstock.com

Inca Funerary Mask of Hammered Gold from Peru

© Marques/Shutterstock.com

Conquistador Francisco Pizarro (d. 1541) on a Spanish Stamp

SEA EXPLORATION AND ITS CONSEQUENCES

SEA EXPLORATION: NOT NEW

- Polynesians explored and settled Eastern Pacific
- Indian Ocean was major center of commerce explored by Chinese + rise of Islam boosted trade in this region
- Vikings, Amerindians, and Africans pursued long-distance exploration
  - Vikings settled Iceland and Greenland (10th c.)
  - Mali Africans explored Atlantic Ocean
  - The transfer of maize cultivation from Mesoamerica to South America included use of boats along Pacific coast

EUROPEAN MARITIME EXPANSION (15TH-16TH C.)

- Surpassed earlier exploration. Why?
- Ended the isolation of the Americas
- Until the 15th c. Europe had remained at the periphery of world history

CAUSES AND MOTIVES

- Revival of urban life and trade in Europe
- Unique financial alliance of merchants and rulers
  - Prince Henry the Navigator opened research center
  - The goal of center was
    - to study navigation and the works of Jewish cartographers
    - to improve navigational tools
- Struggle and competition with Islamic powers for dominance in the Mediterranean Sea
- Growing intellectual curiosity

IMPROVEMENTS IN MARITIME AND MILITARY TECHNOLOGIES

- Magnetic compass first developed in China
- The astrolabe, invented by Greeks and Arabs
- Cannons + guns + steel swords
- New long-distance sailing vessels (the Portuguese caravel)
  - Small (could enter shallow coastal waters and explore upriver)
  - Could withstand strong ocean storms
  - Equipped with triangular lateen sail
  - Addition of small cannons

WHY IBERIA?

- MAIN GOAL: to reach spice-producing lands of South and Southeast Asia by sea to bypass Ottoman Turks who controlled land routes and Eastern Mediterranean Sea
- Compared to Italy, Iberia had a modest share of Mediterranean trade

- Iberia had been engaged in anti-Muslim warfare since 8th c. Christian militancy drove their overseas ventures

IBERIANS AND MONGOLS: THAT DIFFERENT?

- Very similar
  - Warrior culture
  - Used technology adopted from other civilizations
  - Killed thousands through combat, slaughter, and diseases
  - Connected cultures
- Different
  - Iberians zealously imposed their faith on people they conquered

PORTUGAL

- Goals = Africa, India, Indonesia
- Captured ports along Morocco's Atlantic coast
- Explored Western African coast
- Discovered island of Sao Tome on the equator and transformed it into a sugar plantation dependent on slave labor [model for future sugar plantations of Brazil and the Caribbean]

PORTUGAL

- In 1497-1498, Vasco da Gama sailed around Africa and reached India
- In 1500, Pedro Alvares Cabral accidentally reached Brazil
- By the early 16th c. the Portuguese empire stretched from Africa to Indonesia (spice trade)
- In general, they negotiated trading rights in contracts that benefited both local merchants and themselves

SPAIN

- 1492 = Christopher Columbus sailed West and discovered a New World (the Bahamas) instead of the Indies
  - Met peaceful Indians, the Arawaks
  - During second trip, he tried to find the imaginary gold mines but failed + switched to find slaves (enemies of the Arawaks)
- Later, Spaniards forced Indians to work on sugarcane plantations
- After Columbus, Spain colonized Caribbean islands, Aztecs in Mexico, Incas in Peru
- 1522 Ferdinand Magellan sailed around the globe for the first time

AZTECS

- Called themselves the Mexica = spoke Nahuatl
- Capital city Tenochtitlan on an island. Canoes crowded the city
- Former nomads and mercenaries
- When they took power they burned the books of earlier civilizations
- Their empire was multiethnic + extractive (they received rich tribute in food and cotton from conquered people who could retain their own leaders)
- Their society was militaristic, authoritarian, and hierarchical
- Women could inherit property and had access to priestly roles

HUMAN SACRIFICES

- Human sacrifice on a prodigious scale
- Belief was that Sun-god required human blood to sustain him as he battled with the moon and stars
- These sacrifices created resentment, which might explain why many sided with Cortes
- Was this exaggerated by Spaniard writers to justify their conquest of Mexico?
- Maybe, but there is archeological evidence to support their claim

ENCOUNTER WITH SPAIN

- The Spanish possession of gunpowder + diseases reduced the Aztec population from 25 million to 2 million
- Cortes razed their capital city (Tenochtitlan)

INCAS (PERU): ONE OF THE LARGEST EMPIRES IN THE WORLD

- Inca empire = started expansion in 15th c.
- Ethnically and linguistically diverse
- Pachacuti = famous ruler
  - Created one of the world's biggest empires
  - Made Quechua the language of official business
- Developed irrigated agriculture of highland valleys + pottery + urban centers + cotton weaving
- Remarkable system of roads and rope bridges
- Elite exacted taxation in terms of forced labor

INCAS: WRITING AND RELIGION

- The Incas lacked writing but kept detailed accounts using knotted strings (*quipu*)
- Temples dedicated to the sun and moon gleamed with gold and silver
- The Incas saw themselves in a profound relationship to their dead ancestors (mummified bodies of ancestors brought out for special ceremonies)

ENCOUNTER WITH SPAIN

- Francisco Pizarro
- Won partly because he arrived at the time of a family dispute within the Inca imperial family for political power + he allied himself with the Incas' enemies
- Conquests provided Spain with silver

LIVES OF PREVIOUSLY ISOLATED CIVILIZATIONS CHANGED

- Europeans brought their animals, foods, plants, diseases, and religion
- Spaniards had no interest in preserving foreign cultures (Aztec, Inca, Maya)
- Imposed Christianity and destroyed Incas' mummies
- In 16th c., Christian orders such as Franciscans, Dominicans, and later Jesuits, changed methods and set about converting Indians peacefully

WHY DID EARLY ENCOUNTERS BETWEEN EUROPEANS AND AMERICAN INDIANS TAKE SUCH A DIFFERENT TURN (AS COMPARED TO AFRICA)?

AFRICA

- No colonial subjugation of one group by another
- Africans had long access to iron weaponry and used horse cavalry
- Europeans died at alarming rate in Africa (no immunity to malaria and yellow fever)

AMERICAS

- Colonial subjugation of one group by another
- Europeans brought down mighty empires (Why? No iron weaponry, no horse power, no guns)
- Indigenous people fell in great numbers when exposed to pathogens introduced from Afroeurasia

CONSEQUENCES FOR EUROPE

- Regular oceanic trade across vast spaces (between Atlantic and Indian oceans + between West Africa and Brazil + between North and South Atlantic)
  - Spices, silk, sandalwood reached Europe
  - Brazilian tobacco was smoked in Portugal but also Angola
  - Diseases also traveled (malaria + syphilis)
- The Atlantic Ocean soon replaced the Indian Ocean and the Mediterranean Sea as centers of world commerce

COLUMBIAN EXCHANGE

- DEFINITION: Global diffusion of plants, crops, animals, human populations, and disease pathogens after first European explorations
- Syphilis came from America
- In Europe many human diseases originated in livestock: measles, smallpox, influenza. They came to America

SLAVERY

- Travels expanded its scale
- Europeans started attaching race and color to servitude
- From the 15th c. = Africans increasingly filled the ranks of slaves (toiled in sugar plantations)
- American Indians were not enslaved
- In the Americas, slavery flourished as an institution of exploitation

SLAVERY

BEFORE 15TH C.

- Colorblind. Slaves could be Greek, Slav, European, African, and Turk
  - Prisoners of war
  - Slavery for debt
  - Sold by parents
- In Islam, constituted important elite contingent (ex: Janissaries) or administrators
- African slaves worked as domestic slaves or household servants of the aristocracy (considered as prestigious symbols of wealth)

AFTER TRAVELS

- Race and color attached to servitude
- Employed primarily for economic production
- In Americas, slavery = institution of exploitation

CONCLUSION

- Europe increased its dominance through its exploitation of Africa and the New World
- However, the Muslim world in 1500 (gunpowder empires in Turkey, India, and Iran) was still quite strong and could successfully compete with the Portuguese in the Indian Ocean

THE AMERICAS IN 1500

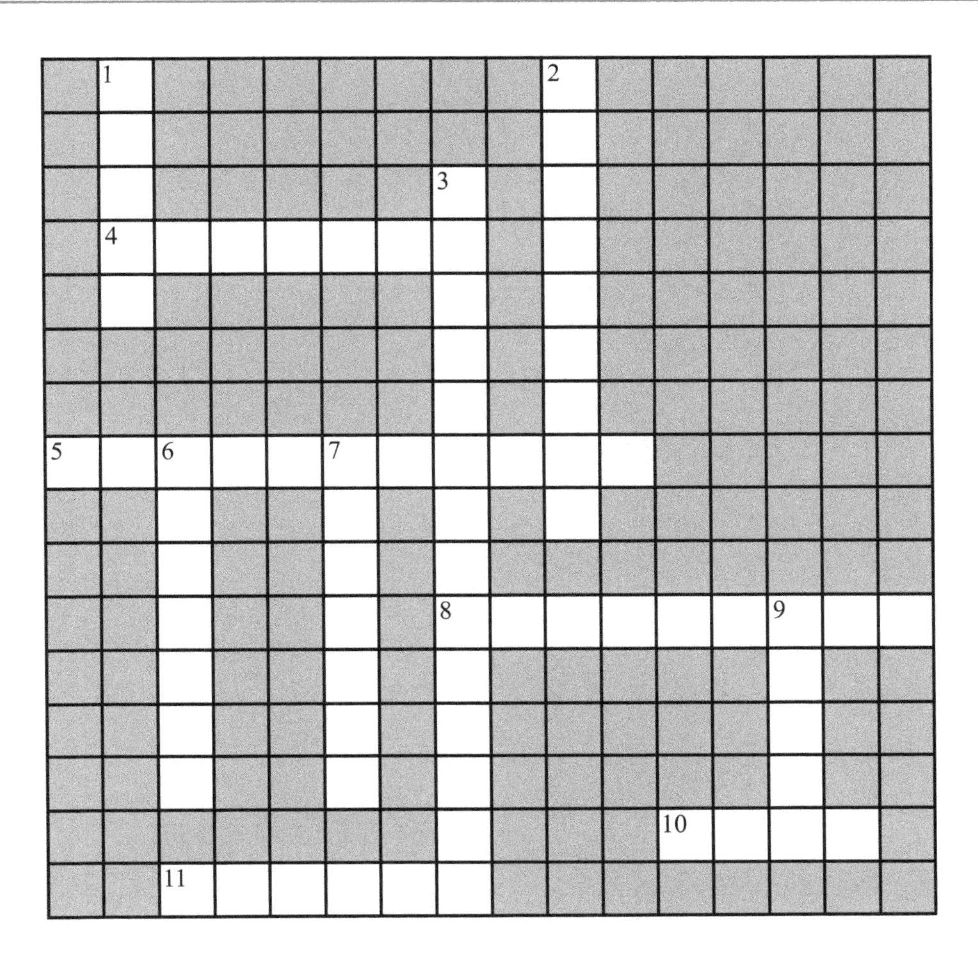

ACROSS

4. Spanish explorer who conquered the Inca empire and founded Lima, the capital of Peru
5. The Christian reconquest of Iberia from the Muslims
8. Navigational instrument for determining latitude invented by Greeks and Arabs
10. Largest and most powerful Andean empire which controlled the Pacific coast of South America from Ecuador to Chile from its capital of Cuzco
11. Spanish commander who conquered the Aztec empire in Central America

DOWN

1. Knotted string used by Andean peoples for record keeping
2. Famous fifteenth-century Inca emperor who created one of the world's biggest empires and made Quechua the language of official business
3. Spanish adventurers such as Cortes and Pizarro who conquered Central and South America in the sixteenth century (Spanish spelling)
6. Small and fast Spanish-Portuguese sailing boat equipped with lateen sail and small cannons
7. Language spoken by the Incas of Peru
9. Central American empire constructed by the Mexica who spoke Nahuatl

Article Twelve

The Real Story of Globalization

BY CHARLES C. MANN

In the great tropical harbor of Manila Bay, two groups of men warily approach each other, their hands poised above their weapons. Cold-eyed, globe-trotting traders, they are from opposite ends of the earth: Spain and China.

The Spaniards have a big cache of silver, mined in the Americas by Indian and African slaves; the Chinese bring a selection of fine silk and porcelain, materials created by advanced processes unknown in Europe. It is the summer of 1571, and this swap of silk for silver--the beginning of an exchange in Manila that would last for almost 250 years--marks the opening salvo in what we now call globalization. It was the first time that Europe, Asia and the Americas were bound together in a single economic network.

The silk would cause a sensation in Spain, as the silver would in China. But the crowds that greeted the returning ships had no idea what they were truly carrying. We usually describe globalization in purely economic terms, but it is also a biological phenomenon. Researchers increasingly think that the most important cargo on these early transoceanic voyages was not silk and silver but an unruly menagerie of plants and animals, many of them accidental stowaways. In the sweep of history, it is this biological side of globalization that may well have the greater impact on the fate of the world's people and nations.

Some 250 million years ago, the Earth contained a single landmass known as Pangaea. Geological forces broke up this vast expanse, forever splitting Eurasia and the Americas. Over time the two halves of Pangaea developed wildly different suites of plants and animals.

Before Columbus sailed the Atlantic, only a few venturesome land creatures, mostly insects and birds, had crossed the oceans and established themselves. Otherwise, the world was sliced into separate ecological domains. Columbus's signal accomplishment was, in the phrase of the historian Alfred W. Crosby, to reknit the seams of Pangaea.

After 1492, the world's ecosystems collided and mixed as European vessels carried thousands of species to new homes across the oceans. The Columbian Exchange, as Mr. Crosby called it, is why we came to have tomatoes in Italy, oranges in Florida, chocolate in Switzerland and chili peppers in Thailand.

A growing number of scholars believe that the ecological transformation set off by Columbus's voyages was one of the establishing events of the modern world. Why did Europe rise to predominance? Why did China, once the richest, most advanced society on earth, fall to its knees? Why did chattel slavery take hold in the Americas? Why was it the United Kingdom that launched the Industrial Revolution? All of these questions are tied in crucial ways to the Columbian Exchange.

Where to start? Perhaps with the worms. Earthworms, to be precise--especially the common nightcrawler and the red marsh worm, creatures that did not exist in North America before 1492.

Well before the start of the silk-and-silver trade across the Pacific, Spanish and Portuguese conquistadors were sailing the Atlantic in search of precious metals. Ultimately, they exported huge supplies of gold and silver from Bolivia, Brazil, Colombia and Mexico, vastly increasing Europe's money supply. But those homebound ships contained something else of equal importance: the Amazonian plant known today as tobacco.

Intoxicating and addictive, tobacco became the subject of the first truly global commodity craze. By 1607, when England founded its first colony in Virginia, London already had more than 7,000 tobacco "houses"--cafe-like places where the city's growing throng of nicotine junkies could buy and smoke tobacco. To feed the demand, English ships tied up to Virginia docks and took in barrels of rolled-up tobacco leaves. Typically 4 feet tall and 21/2 feet across, each barrel weighed half a ton or more. Sailors balanced out the weight by leaving behind their ships' ballast: stones, gravel and soil. They swapped English dirt for Virginia tobacco.

That dirt very likely contained the common nightcrawler and the red marsh worm. So, almost certainly, did the rootballs of plants that the colonists imported. Before Europeans arrived, the upper Midwest, New England and all of Canada had no earthworms--they had been wiped out in the last Ice Age.

In worm-free woodlands, leaves pile up in drifts on the forest floor. Trees and shrubs in wormless places depend on litter for food. When earthworms arrive, they quickly consume the leaf litter, packing the nutrients deep in the soil in the form of castings (worm excrement). Suddenly, the plants can no longer feed themselves; their fine, surface-level root systems are in the wrong place. Wild sarsaparilla, wild oats, Solomon's seal and a host of understory plants die off; grass-like species such as Pennsylvania sedge take over. Sugar maples almost stop growing, and ash seedlings start to thrive.

Spread today by farmers, gardeners and anglers, earthworms are obsessive underground engineers, and they are now remaking swathes of Minnesota, Alberta and Ontario. Nobody knows what will happen next in what ecologists see as a gigantic, unplanned, centuries-long experiment.

Before Columbus, the parasites that cause malaria were rampant in Eurasia and Africa but unknown in the Americas. Transported in the bodies of sailors, malaria may have crossed the ocean as early as Columbus's second voyage. Yellow fever, malaria's frequent companion, soon followed.

By the 17th century, the zone where these diseases held sway--coastal areas roughly from Washington, D.C., to the Brazil-Ecuador border--was dangerous territory for European migrants, many of whom died within months of arrival. By contrast, most West Africans had built-in defenses, acquired or genetic, against the diseases.

Initially, American planters preferred to pay to import European laborers--they spoke the same language and knew European farming methods. They also cost less than slaves bought from Africa, but they were far less hardy and thus a riskier investment. In purely economic terms, the historian Philip Curtin has calculated, the diseases of the Columbian Exchange made the enslaved worker "preferable at anything up to three times the price of the European."

Did the Columbian Exchange cause chattel slavery in the Americas? No. People are moral agents who weigh many considerations. But anyone who knows how markets work will understand the pull exerted by slavery's superior profitability.

Much more direct was the role of the Columbian Exchange in the creation of Great Britain. In 1698, a visionary huckster named William Paterson persuaded wealthy Scots to invest as much as half the nation's available capital in a scheme to colonize Panama, hoping to control the chokepoint for trade between the Pacific and the Atlantic. As the historian J.R. McNeill recounted in "Mosquito Empires," malaria and yellow fever quickly slew almost 90% of the 2,500 colonists. The debacle caused a financial meltdown.

At the time, England and Scotland shared a monarch but remained separate nations. England, the bigger partner, had been pushing a complete merger for decades. Scots had resisted, fearing a London-dominated economy, but now England promised to reimburse investors in the failed Panama project as part of a union agreement. As Mr. McNeill wrote, "Thus Great Britain was born, with assistance from the fevers of Panama."

But Scots could hardly complain about the consequences of the Columbian Exchange. By the time they were absorbed into Britain, their daily bread, so to speak, was a South American tuber now familiar as the domestic potato.

Compared with grains, tubers are inherently more productive. If the head of a wheat or rice plant grows too big, the plant will fall over, killing it. There are no structural worries with tubers, which grow underground. Eighteenth-century farmers who planted potatoes reaped about four times as much dry food matter as they did from wheat or barley.

Hunger was then a familiar presence in Europe. France had 40 nationwide food calamities between 1500 and 1800, more than one every decade, according to the French historian Fernand Braudel. England had still more. The continent simply could not sustain itself.

The potato allowed most of Europe--a 2,000-mile band between Ireland and the Ukraine--to feed itself. (Corn, another American crop, played a similar role in Italy and Romania.) Political stability, higher incomes and a population boom were the result. Imported from Peru, the potato became the fuel for the rise of Europe.

The sweet potato played a similar role in China. Introduced (along with corn) from South America via the Pacific silver trade in the 1590s, it suddenly provided a way for Chinese farmers to cultivate upland areas that had been unusable for rice paddies. The nutritious new crop encouraged the fertility boom of the Qing dynasty, but the experiment soon went badly wrong.

Because Chinese farmers had never cultivated their dry uplands, they made beginners' mistakes. An increase in erosion led to extraordinary levels of flooding, which in turn fed popular unrest and destabilized the government. The new crops that had helped to strengthen Europe were a key factor in weakening China.

The Columbian Exchange carried other costs as well. When Spanish colonists in Hispaniola imported African plantains in 1516, the Harvard entomologist Edward O. Wilson has proposed, they also brought over some of the plant's parasites: scale insects, which suck the juices from banana roots.

In Hispaniola, Mr. Wilson argues, these insects had no natural enemies. Their numbers must have exploded--a phenomenon known as "ecological release." The spread of scale insects would have delighted one of the region's native species: the tropical fire ant, which is fond of dining on the sugary excrement of scale insects. A big increase in scale insects would have led to a big increase in fire ants.

This is only informed speculation. What happened in 1518 and 1519 is not. According to an account by a priest who witnessed those years, Spanish homes and plantations in Hispaniola were invaded by "an infinite number of ants," their stings causing "greater pains than wasps that bite and hurt men." Overwhelmed by the onslaught, Spaniards abandoned their homes to the insects, depopulating Santo Domingo. It was the first modern eco-catastrophe.

A second, much more consequential disaster occurred two centuries later, when European ships accidentally imported the fungus-like organism, native to Peru, that causes the potato disease known as late blight. First appearing in Flanders in June 1845, it was carried by winds to potato farms around Paris in August. Weeks later it wiped out fields in the Netherlands, Germany, Denmark and England. Blight appeared in Ireland on Sept. 13.

The Irish were more dependent on potatoes than any other Western nation. Within two years, more than a million died. Millions more fled. The nation never regained its footing. Today Ireland has the melancholy distinction of being the only nation in Europe, and perhaps the world, to have fewer people within the same boundaries than it did more than 150 years ago.

The Columbian Exchange continues to this day. The Pará rubber tree, originally from Brazil, now occupies huge swathes of southeast Asia, providing the latex necessary to make the tires, belts, O-rings and gaskets that invisibly maintain industrial civilization. (Synthetic rubber of equal quality still cannot be practicably manufactured.)

Asian rubber plantations owe their existence to a British swashbuckler named Henry Wickham, who in 1876 smuggled 70,000 rubber seeds from Brazil to London's Kew Gardens. Rubber-tree plantations are next to impossible in the tree's Amazonian home, because they are wiped out by an aggressive native fungus, Microcyclus ulei. Much as the potato blight crossed the Atlantic, M. ulei will surely make its way across the Pacific one day, with consequences as disastrous as they are predictable.

Species have always moved around, taking advantage of happenstance or favorable circumstances. But the Columbian Exchange, like a biological Internet, has put every part of the natural world in contact with every other, refashioning it, for better or worse, at a staggering rate.

The consequences are as hard to predict as those of globalization itself. Even as plantations of Brazilian rubber take over tropical forests in Southeast Asia, plantations of soybeans, a Chinese legume, are replacing almost 80,000 square miles of the southern Amazon, an area almost the size of Britain. In dry northeastern Brazil, Australian eucalyptus covers more than 15,000 square miles. Returning the favor, entrepreneurs in Australia are now attempting to establish plantations of açaí, a Brazilian palm tree whose fruit has been endorsed by celebrities as being super-healthful.

All of these developments will yield positive economic results--soy exports, for instance, are making Brazil into an agricultural powerhouse, lifting the fortunes of countless poor farmers in remote places. But the downside of the ongoing Columbian Exchange is equally stark. Forests in the U.S. are being devastated by a host of foreign pests, including sudden oak death, a cousin of potato blight that is probably from southern China; the emerald ash borer, an insect from northern China that probably arrived in ship pallets; and white pine blister rust, a native of Siberia first seen in the Pacific Northwest in 1920.

Forests full of dead trees are prone to catastrophic fires, a convulsive agent of change. New species will rush in to replace those that are lost, with effects that cannot be known in advance. We will simply have to wait to see what kind of landscape our children will inherit.

Today our news is dominated by stories of debt deals and novel computer applications and strife in the Middle East. But centuries from now, historians may well see our own era as we have started to see the rise of the modern West: as yet another chapter in the unfolding tumult of the Columbian Exchange.

After reading article 12, define the term "Columbian exchange" and discuss the effects of the exchange using examples of three different species.

Selected References

Bentley, Jerry H., Herbert F. Ziegler, and Heather E. Streets-Salter. *Traditions and Encounters: A Global Perspective on the Past.* New York: McGraw Hill, 2015.

Bulliet, Richard, Pamela Kyle Crossley, Daniel R. Headrick, Steven W. Hirsch, Lyman L. Johnson, and David Northrup. *The Earth and Its Peoples: A Global History*. Stamford, CT: Cengage Learning, 2015.

Coatsworth, John, Juan Cole, Michael P. Hanagan, Peter C. Perdue, Charles Tilly, and Louise Tilly. *Global Connections: Politics, Exchange, and Social Life in World History*. Cambridge: Cambridge University Press, 2015.

Craig, Albert M., William A. Graham, Donald Kagan, Steven Ozment, and Frank M. Turner. *The Heritage of World Civilizations*. Boston: Prentice Hall, 2011.

Dunn, Ross E., and Laura J. Mitchell. *Panorama: A World History*. New York: McGraw Hill, 2015.

Esposito, John L. *Islam: The Straight Path*. New York: Oxford University Press, 1998.

Esposito, John, Darrell J. Fasching, and Todd Lewis. *World Religions Today*. New York: Oxford University Press, 2006.

Fernández-Armesto, Felipe. *The World: A History*. Vol. 1. Upper Saddle River, NJ: Prentice Hall, 2010.

Hollister, C. Warren, J. Sears McGee, and Gale Stokes. *The West Transformed: A History of Western Civilization.* Fort Worth, TX: Harcourt College Publishers, 2000.

Judge, Edward H., and John W. Langdon. *Connections: A World History.* Boston: Pearson, 2016.

McCarthy, Justin. *The Ottoman Turks: An Introductory History to 1923.* London and New York: Longman, 1997.

Molloy, Michael. *Experiencing the World's Religions: Tradition, Challenge, and Change*. New York: McGraw Hill, 2010.

Nielsen, Neils C., Jr., Norvin Hein, Frank E. Reynolds, et al. *Religions of the World*. New York: St. Martin's Press, 1993.

Noss, David S. *A History of the World's Religions.* Upper Saddle River, NJ: Prentice Hall, 2003.

Oxtoby, Willard. *World Religions: Western Traditions.* Ontario: Oxford University Press, 2002.

Schmidt, Roger, Gene C. Sager, Gerald T. Carney, Albert Charles Muller, Kenneth J. Zanca, Julius J. Jackson, Jr., C. Wayne Mayhall, and Jeffrey C. Burke. *Patterns of Religion*. Belmont, CA: Thompson Wadsworth, 2005.

von Sivers, Peter, Charles A. Desnoyers, and George B. Stow. *Patterns of World History*. Vol. 1. New York: Oxford University Press, 2015.

Young, William. *The World's Religions: Worldviews and Contemporary Issues.* Upper Saddle River, NJ: Prentice Hall, 2005.

CPSIA information can be obtained
at www.ICGtesting.com
Printed in the USA
LVHW051338290622
722246LV00002B/7